# THE ONCE AND FUTURE SILVER QUEEN OF THE ROCKIES

## Mining the American West

**Series Editors:** DUANE A. SMITH • ROBERT A. TRENNERT • LIPING ZHU

*Boomtown Blues: Colorado Oil Shale*
ANDREW GULLIFORD

*From Redstone to Ludlow: John Cleveland Osgood's Struggle against the United Mine Workers of America*
F. DARRELL MUNSELL

*Gambling on Ore: The Nature of Metal Mining in the United States, 1860–1910,*
KENT A. CURTIS

*Hard as the Rock Itself: Place and Identity in the American Mining Town*
DAVID ROBERTSON

*High Altitude Energy: A History of Fossil Fuels in Colorado*
LEE SCAMEHORN

*A History of Gold Dredging in Idaho*
CLARK C. SPENCE

*Industrializing the Rockies: Growth, Competition, and Turmoil in the Coalfields of Colorado and Wyoming*
DAVID A. WOLFF

*The Mechanics of Optimism: Mining Companies, Technology, and the Hot Spring Gold Rush, Montana Territory, 1864–1868*
JEFFREY J. SAFFORD

*Mercury and the Making of California: Mining, Landscape, and Race, 1840–1890,*
ANDREW SCOTT JOHNSTON

*The Once and Future Silver Queen of the Rockies: Georgetown, Colorado, and the Fight for Survival into the Twentieth Century*
CHRISTINE A. BRADLEY AND DUANE A. SMITH

*The Rise of the Silver Queen: Georgetown, Colorado, 1859–1896*
LISTON E. LEYENDECKER, DUANE A. SMITH, AND CHRISTINE A. BRADLEY

*Santa Rita del Cobre: A Copper Mining Community in New Mexico*
CHRISTOPHER J. HUGGARD AND TERRENCE M. HUMBLE

*Silver Saga: The Story of Caribou, Colorado, Revised Edition*
DUANE A. SMITH

*Thomas F. Walsh: Progressive Businessman and Colorado Mining Tycoon*
JOHN STEWART

*Yellowcake Towns: Uranium Mining Communities in the American West*
MICHAEL A. AMUNDSON

# THE ONCE AND FUTURE SILVER QUEEN OF THE ROCKIES

## GEORGETOWN, COLORADO, AND THE FIGHT FOR SURVIVAL INTO THE TWENTIETH CENTURY

**Christine A. Bradley and Duane A. Smith**

UNIVERSITY PRESS OF COLORADO
*Louisville*

Published by University Press of Colorado
245 Century Circle, Suite 202
Louisville, Colorado 80027


Printed in the United States of America

ASSOCIATION of UNIVERSITY PRESSES The University Press of Colorado is a proud member of the Association of University Presses.

The University Press of Colorado is a cooperative publishing enterprise supported, in part, by Adams State University, Colorado State University, Fort Lewis College, Metropolitan State University of Denver, University of Colorado, University of Northern Colorado, Utah State University, and Western State Colorado University.

∞ This paper meets the requirements of the ANSI/NISO Z39.48-1992 (Permanence of Paper).

ISBN: 978-1-60732-607-6 (paperback)
ISBN: 978-1-60732-608-3 (ebook)
DOI: https://doi.org/10.5876/9781607326083

Library of Congress Cataloging-in-Publication Data

Names: Bradley, Christine A. (Christine Ann), 1952– author. | Smith, Duane A., author.
Title: The once and future Silver Queen of the Rockies : Georgetown, Colorado and the fight for survival into the 20th century / Christine A. Bradley and Duane A. Smith.
Other titles: Georgetown, Colorado and the fight for survival into the 20th century | Mining the American West.
Description: Boulder : University Press of Colorado, [2017] | Series: Mining the American West series | Includes bibliographical references and index.
Identifiers: LCCN 2017033392 | ISBN 9781607326076 (pbk.) | ISBN 9781607326083 (ebook)
Subjects: LCSH: Georgetown (Colo.)—History—19th century. | Georgetown (Colo.)—History—20th century. | Silver mines and mining—Colorado—Georgetown—History—Pictorial works. | Gold mines and mining—Colorado—Georgetown—History—Pictorial works. | Tourism—Colorado—Georgetown. | Historic preservation—Colorado—Georgetown.
Classification: LCC F784.G34 B73 2017 | DDC 978.8/61—dc23
LC record available at https://lccn.loc.gov/2017033392

# CONTENTS

# PREFACE

*The Once and Future Silver Queen* documents the struggles of Georgetown, Colorado, as it turned from a volatile economy based upon silver mining to the demands of a permanent town without much to support it. Tourism, government, and historic preservation filled the void; the tiny town in the Rockies found a way to survive. The exuberance of the nineteenth century faded, yet the town managed to survive and retain enough of its past (probably more through a lack of alternatives than a conscious choice) so that people could look around in the early 1970s and appreciate a treasure trove of nineteenth-century architecture in a lovely setting. Stately homes, miners' cabins, and a small commercial district became the basis for a new economy.

Christmas eve, 1969, Ronald J. Neely, John Calhoun, and Cynthia Wadsworth (later Neely) sat in front of the fireplace at the Alpine Inn with glasses of wine and an occasional cognac, wondering about the future of the town. Threats of development needed to balance with economic survival. As the discussion picked up speed, a dream started to emerge. They drank a toast to "The once and future Silver Queen" as they realized the town had survived and could begin to regain its former splendor—not as a mining community, but as a historic gem.

Over the next four decades, Ron would be the skipper of the ship. He would tell you he had lots of help—we would tell you it wasn't always happy sailing—and while we look back with pride, we still shake our heads and wonder how all of it came to be.

This book is dedicated to Ron, the dream weaver, whose ability to ignore fiscal reality and plow forward with nothing more than a belief in his community and its people, made many things possible. The story documents the transition from the old Silver Queen to the new—up to the start of the historic preservation era in Georgetown in the early 1970s. While there is a definite interest in the story of historic preservation in Georgetown, that story will require some "distance" and a perspective more unbiased than my own.

CHRISTINE A. BRADLEY
*Georgetown, Colorado*

# THE ONCE AND FUTURE SILVER QUEEN OF THE ROCKIES

# CHAPTER 1

# DAYS OF HOPE, YEARS OF DESPAIR

"Marley was dead, to begin with, there is no doubt whatever about that." Charles Dickens penned those opening words over a half century before Christmas eve, 1896. His *Christmas Carol* had been published long before Georgetown emerged on the silver scene. As Georgetown folk prepared now for Christmas, in that depressing year, 1896, they might have feared that they, and their era, were as dead as what Dickens had written about Scrooge's partner. "Old Marley was as dead as a door-nail."

While Christmas Eve darkened over Georgetown and the valley, this favorite Christmas tale was probably being read aloud to families gathered around the fireplace or potbellied stove in countless homes. As the story unfolded, the opening might have struck a receptive chord about the situation for the dejected locals. Bleak and cold as Dickens's London was on that long-ago foggy Christmas Eve, Georgetown seemed almost as "cold, bleak," and "biting."

The year 1896 had been filled with contrasting emotions of hope and expectation, reality and despair. More of the latter than the former, filled

Christmas stockings that season, as Georgetown, Silver Plume, and the silver-mining West suffered through a few highs but mostly lows.

The gloom generated by the crash of 1893 was compounded by the gripping Depression that had descended on the area that summer. How bad was it? The Colorado Bureau of Labor Statistics asked "prominent citizens" of each county to comment on the situation in the fall of 1893. Clear Creek's respondent starkly stated, when dealing with the Georgetown area, "It is a very serious question with us now whether they [silver mines] will ever start up again, as the outlook at present is very blue."[1]

Certainly, circumstances would not worsen, but every year it seemed to, as the darkening depression settled dolefully over town, county, and state. The recently lost 1896 presidential election only made matters more emotional and more disheartening. William McKinley and gold were swept into office to reign triumphantly. He was not the hero of the Georgetown silverites. Their hero, William Jennings Bryan, had lost along with desperately desired "Free Silver," with its promised revival of a higher silver price. These had been a last hope as the savior for disheartened local mining. Georgetown had done its best, voting over 30 to 1 for the "silver tongued orator."

The *Georgetown Courier*'s editor bitterly remarked in the November 7, 1896, issue: "We have no doubt that there is a remarkable coincidence in the number of McKinley votes cast in Georgetown and the number of applicants for the post office."

Marley might not be dead in Georgetown, but the residents' beloved "free silver" cause was, even if they futilely wished not to face that reality. Substituting but two words for one, their plight was told by Dickens: "There is no doubt that [free silver] was dead. This must be distinctly understood, or nothing wonderful can come of the story." For Georgetown folks, nothing wonderful would come tomorrow or in the tomorrows stretching into the future.

The Georgetown paper realized this, even though some of its readers could not bring themselves to agree: "For three years the people of Georgetown and other silver mining camps have been down to hard-pan. We did not start in at bed-rock on a four years' crusade with nothing in sight and with no hope of finding anything. Heaven temper the wind of an impoverished people."[2] The opening years of the 1890s had not been good to Georgetown and its chief industry. The price of silver had dropped from an 1890 high of a $1.05 per ounce to 68 cents in 1896, and that referred to smelted and refined silver, not to the raw ore coming out of the mine portal. Coloradans feared that the price might drop lower, a not unreasonable fear considering the times and the world monetary market.

To make matters worse, the district was over thirty years old, too old for a mining region. The seasons of its youth withered into only a fading memory, already vanishing into legend.

The cost of mining had gone up as the mines went down, and, as part of a disheartening general pattern, the value of the ore decreased with depth. Investors looked to new, exciting districts, ones with seemingly unlimited potential, not ones where the best locations and richest deposits had long ago been staked and worked for a generation. The future would not have been all that rosy, even without the trials and tribulations of the 1890s.

Like their counterparts elsewhere, Georgetown's miners and the other folks did not give up hope. They believed that the federal government would eventually see the necessity of putting silver back on a 16 to 1 ratio with gold. With gold stabilized at $20 plus change, silver would be at $1.25, a price last seen in 1875. Those had been flush times, and Georgetown thirsted to return to the "good old days."

The New Year did not dawn with a brighter tomorrow. Georgetown and its silver mines were locked virtually in a life-and-death struggle to survive. Hope could only carry them so far before it would die "a starving," as Ben Franklin had reminded his readers well over a century before. But with a never-say-die spirit, which was a western tradition, Georgetown set out to reestablish its silver foundation and push onward to better times in the upcoming new century.

As the *Georgetown Courier* proclaimed on its front page:

> The Oldest newspaper in Clear Creek County
>
> OUR INDUSTRY, MINING
>
> this district has produced over $60,000,000

It was not that the district suddenly shut down during the 1890s. The generosity of Congress in 1890 in passing the Sherman Silver Purchase Act—and the corresponding spike in the price of silver with increased government purchase—created $1.3 million in silver production that year. The county did not drop under the million-dollar production level until 1897, the last year it reached that plateau.

A check of the *Engineering and Mining Journal*'s and the *Courier*'s mining columns and notes reveals the times and the trends, as Georgetown struggled to survive in a seemingly hostile environment.

> The Colorado Central Mine once a large steady producer apparently has been ruined by long and costly litigation. It will probably be added to the list of abandoned mines in this district.[3]
>
> Prospectors recently opened gold bearing ore at the Cyclone on McClellan Mt. near Silver Plume. They were doing assessment work and opened a vein of 7 ounce gold.[4]
>
> The western end of Clear Creek County is one of the very few mining districts in Colorado working out its own salvation. For several years capital fought shy of the district because its product is mainly silver. Yet for all that with the low price of silver and lead the product has averaged about $1,000 for each man engaged in mining.[5]
>
> It is infinitely easier for a needle to go through the eye of a camel than for a local paper to gather mining information in this district. Mining here is generally regarded as a private enterprise in which the public has no business.[6]
>
> Georgetown has some silver mines working. The town has never recovered from the great depression.[7]

In many ways, the last two excerpts are the most revealing. The *Courier* carried less mining news in these years than Georgetown papers had ever had before. It did carry news on the Idaho Springs District, as did the *Engineering and Mining Journal*. That gold district was going through a revival, as the county's gold production stayed steady—in the high $600,000 to $700,000 range—and eventually surpassed its longtime up-county silver rival. As might be imagined, silver production actually improved somewhat over previous totals, even as miners and investors turned their attention to the more profitable golden mineral. Silver as a by-product was sometimes found in gold mines.

The secretiveness is interesting. Usually camps and mines in bonanza did not mind a little publicity. If nothing else, it helped tempt investors, those individuals who were usually most welcome in any district. Georgetown owners apparently did not want such notoriety. Perchance, however, they had very little to point to with much pride or potential.

Mines ruined by litigation did not bode well for the district. Gold accidentally found while doing assessment work was a rare exception and not a pillar on which to build the future. Blaming easterners for western problems was as old as the hills, and there was definitely truth in the statement that Georgetown had never recovered.

The *Courier* indicated all this when it carried more mining news from Idaho Springs than it did from Georgetown: "Eastern people have recently secured property on Bellview Mountain. They are spending considerable money on development without having made a showing. The men in charge do not appear to be getting work out of the investment. It is for this reason and no other that there are failures in that section on properties which show large bodies of high grade ore."[8]

Development in some mines had been hurt by poor management, and, in reality, only yearly assessment work was being done. Fortunately in one case, and purely by accident, miners discovered gold. On nonpatented claims (ones that were not owned outright or patented), federal law mandated that at least $100 worth of assessment work be done every year before December 30 in order to hold the claim against relocation by other parties.[9]

The fact that the claims were nonpatented and that only assessment work was conducted told a story. Owners still held to the hope that they might discover valuable ore but did not have the confidence, or money, to put that idea to the test beyond just maintaining ownership. Mining was becoming more expensive every year and the odds of finding a bonanza decreasing.

All the major mines had been patented, which included $500 worth of "patent improvement," including "officially" surveying the claim. It meant advertising the application in a local newspaper for the benefit of anyone who might have an adverse claim on the property and, finally, making a written patent application. When all that was accomplished, the government approved the purchase of the land for $5 an acre. With ownership in hand, the property became subject to taxes, but assessment work no longer had to be done each year.

Old as a mining district, Clear Creek County showed the disparity between patented and nonpatented claims. The 1902 report of the Colorado Bureau of Mines noted "nearly 23,000 lode and placer claims, mill and tunnel sites duly recorded."[10] Of those, only 1,300 owners procured a patent.

Another thing that happened in the Georgetown district, and other mining districts fallen on similar hard times, was that the county levied tax liens on mining properties because the owner had failed to pay taxes. If no back taxes were paid, the property could eventually be sold at a tax sale.

Those all flashed warning signs to potential investors. Meanwhile, upper Clear Creek County needed a fresh transfusion of capital. These signs indicated that mining had slipped into a downward cycle, one that would unquestionably prove hard to reverse unless some new and exciting discoveries were found in the surrounding mountains. That was precisely what Silver Plume and Georgetown townspeople wanted, hoped for, and prayed would happen.

They did not sit around and wait. They were determined, despite the low price of silver and apparently no help coming from Washington, to revive their fortunes. That they had looked to the federal government was not unusual. Despite the legendary rugged independence of westerners, from cowboy to miner, Washington had been a silent and sometimes-not-so-silent partner in the opening and settlement of the region. Miners everywhere had turned to Uncle Sam, and their uncle had been very good to them in years past. Under the 1872 Mining Law, for instance, patented mining claims, as mentioned, could be purchased for $5 per acre, an astounding bargain. The mining law itself, which western senators had been crucial in getting passed, had done everything it could to encourage the development of mining.[11] It benefited the whole country to have the West developed, or so those westerners ardently believed and argued.

Furthermore, the federal government encouraged and helped railroads, a vital ingredient in mining's success throughout the West, and the federal army aided in settlement and development. Without Uncle Sam, mining would have been a much more difficult proposition, as would any other attempt to settle the West.

With steadfast determination, upper Clear Creek County refused to give in to the decline in mining, the industry that had given it birth, development, and fame. Silver would rise, and Clear Creek would rise with it. Mining had never actually stopped, and not all of the county's mines experienced the problems previously mentioned.

In order to make the mines more profitable, different techniques had to be tried. Since the start of mining in the district, its high elevation presented problems. So had the cost of mining through shafts, which necessitated lifting ore to the surface and pumping out water the same way. It had not been long before the idea of tunneling in from a lower elevation to intersect the vein gained popularity. This eliminated the expense of hoisting ore to the surface and then transporting it down the mountain. It lowered the cost of hauling supplies and men into the mine while easing the price of dewatering the mine. All these benefits would, in theory, make the more plentiful, lower-grade ore more profitable.

Of course, it was easier said than done, a great deal easier. The initial outlay of tunneling into the vein proved both time consuming and costly. It ran the risk of hitting ore too low in grade to work profitably. As the old mining saying went, "no one knew what was beyond the pick at the end of the mine." With even the most careful planning, it was also possible to trespass on someone else's property underground, which could lead to costly litigation. Therefore, it was better to consolidate a whole group of properties that might be impacted by a tunnel.

These problems had not deterred miners before and would not now. There had been "tunnel crazes" earlier, and the upper Clear Creek District was about to succumb once more, partly through renewed hope and partly through desperation.

The editor of Silver Plume's *Silver Standard* expressed it well when quoting the Georgetown correspondent of Denver's *Rocky Mountain News*: "One of the most formidable undertakings ever agitated in this section is now under consideration. A syndicate of capitalists have been investigating our resources, etc., and will in all probability conclude to prosecute a tunnel into the innermost recesses of Leavenworth and McClellan mountains, starting at a point near the base of Leavenworth mountains at the south end of Georgetown, with the view of intersection with a large number of well-known mines at great depth."[12]

Not finished with such exciting news, the editor continued: "It is thought that a large vein will be encountered by this projected tunnel, also that a large number of undiscovered veins will be disclosed as the tunnel proceeds." There lay, once more, the dreams that had carried mining for eons past, a "large vein" and "undiscovered veins."

This one did not pan out, but that did not deter the optimistic, or those who wanted to mine the investors' pockets through such glittering possibilities. Denver's *Weekly Republican*, January 1, 1900, in its review of mining in the past year, heralded "CLEAR CREEK IS HOME OF TUNNELS." Not finished with such exciting news, the subheadline raised hopes even further: "Bores penetrating deep into the rich treasuries of the mountains—increase in gold production—valuable new discoveries." Then the article opened more promising expectations. "Thousands more miners will find work when the great tunnel enterprises reach the veins of ore which traverse the mountains." It sounded like the boom days all over again.

While the gold discoveries were mainly in the lower part of the county, gold was also being mined in upper Clear Creek. Georgetown and Silver Plume old-timers and newcomers had not lost faith. They saw a better future coming. The article continued, "the citizens of these camps are confident their ores will command a better price in the future, but until that time do not hope for outside investment in the silver mines."

Georgetown would gain the most from all this, the reporter went on to say. It was the center of the ore-buying and -smelting concerns and transportation—and even "many of the miners working in Silver Plume live in Georgetown." Despite some pretense advanced by other camps in the district, Georgetown had always been the dominant mining community and would continue to be so.

A second plan was to have more mines owned and operated by local people, a long spoken and unspoken ideal. They would know local conditions better, be less likely to gut a mine for quick profits, and employ their neighbors. Optimistically, too, they simply would not abandon the property if it did not live up to what they initially expected. Not to mention locals, it was hoped, would possess a higher rate of "staying power."

Instead of relying on outside investors, let the home folks do it. That was the theme of an article that discussed Georgetown, but it was just as applicable to the entire district.

> A home pool is being formed here to take hold of some promising mines which are now lying idle and get a bond and lease and develop them . . . the plan is to form a pool comprising some of our leading business men, limiting the pool to thirty members who will each put up $50, when matter will be pushed forward by putting four men at work on whatever property the board of managers may elect. This will be a good start on this line and will no doubt be the forerunner of several more home pools which will be formed.[13]

Tunnels and home investment pools were proposed to reenergize local mining, but Georgetown and Silver Plume folk were not finished by a long shot. What really was needed, many people thought, were better mills and smelters, with an economical process to reduce the ores to gold and silver.

To help gain these improvements, the "leading citizens of Georgetown and vicinity" organized the Georgetown Board of Mines and Commerce. Its purpose, simply stated, focused on calling "attention to the outside public of the splendid opportunities for investment afforded in our district." The board promised to provide "conservative, reliable information" and encouraged "and invited" investigation.

Not a step behind, Silver Plume organized a Board of Mines and Trade. Likewise, it promised to call attention to the mineral resources and send out printed matter "giving statistics of production and so forth."

To old-timers, all this echoed earlier cries and efforts. In fact, the *Courier* (February 10, 1906) reviewed the history, in a long story: "Georgetown and Its Early Mills and Smelters."[14] The article concluded by discussing Georgetown's greatest prosperity, when the town "had a population of nearly 5,000." For those who had not lived there at the time, it was a pretty dismal repetition of problems, fires, and failures, intermingled with a few successes. Revisionist history was already taking place. Georgetown's population never came close to that many people, according to the census statistics.

Uncle Sam pitched in to help support the district with professional paper number 63 of the United States Geological Survey. The publication of the *Economic Geology of the Georgetown Quadrangle (Together with the Empire District), Colorado* in 1908 provided a detailed look at the area's mining and geology.[15] While not a nighttime reading companion, it nevertheless provided miners and investors with a fascinating look at the past and present.

With 422 pages crammed with maps, drawings, and photographs, it provided a sweeping look at the region, focusing on mine after mine. This was no quickly produced puff publication or self-promoting effort. The volume offered a scientific examination of mining and a variety of related topics. Readers could, for example, select a mine or group of lodes, perhaps the Lebanon group of lodes, which the authors evaluated as having "medium importance." The overall vein structure and mineralogy were then discussed, along with several individual mines, including maps of the principal underground workings with a smidgen of history.

By carefully reading the appropriate pages, the investor would come away with a detailed, professional view of a potential investment. From there, he or she would have some basis to make a judgment independent of local sources. The conclusion might not be as optimistic as the owner would want, but it would be straightforward.

Optimism, however, was not in short supply at the turn of the century, and, when it came to a new mill or process, it proved epidemic. A few examples will suffice.

> The old Clear Creek mill, which has done service for well nigh thirty years, is rapidly going under a change under the supervision of Wm. Rogers. Nearly all the old machinery has been torn out and the interior of the building is being overhauled, and new floors, braces and supports are being put in and the structure generally made as good as new. When this work is completed, the placing of new machinery will be commenced . . . When completed, the mill will be one of the most complete in the country.[16]

> The mill being erected by the Griffith Mines company is fast nearing completion. The greater portion of the machinery has arrived and the contractors are employing a good force of men in the setting and adjusting of the equipment.[17]

> A large force of carpenters are engaged in the construction of the mill [Capital Prize], which is being erected to treat the vast amount of concentrating material that had already been exposed, and the building will soon be ready for the reception of the machinery, which has been seen selected after making exhaustive tests for determining the best method.[18]

> W. E. Malm, in charge of construction and installation of equipment at the Malm chemical-electro plant in the east part of town, has had a force of men at work during the past several days in grading for the switch that is to connect with the Colorado & Southern tracks.[19]

These efforts did not quite match the "process mania" of a generation before, but expectations soared high with every new process and mill.

The plans, construction, machinery, process, and expectations of a new mill held great interest for the general public (perhaps for potential investors also), or so assumed the local editors. They covered these developments in careful detail and, then, as expectations perhaps failed to be met, shifted their attention to some other savior. The key to this anticipation and profit came in working the more abundant low-grade ore in a profitable manner.

One example will suffice, and it is a classic one. The *Courier* breathlessly announced in its March 21, 1914, edition, "The Malm Mill to Be Rushed to Completion." In the coming summer, another article predicted they "will see the Malm dry chlorination works of Georgetown in active operation." The original plant had been commenced in 1909, "and $100,000 spent in the erection of a building, the construction of cement foundations, and the installation of machinery, etc."

Work had stopped because of the "stringency of the money market" before the project as finished. Now John L. Malm, an electrochemist who had "spent years in research and experiment in developing this process," rode to the rescue. He had worked out the "bugs" in an experimental plant built in Corbin, Montana.

Georgetownites and others in the upper Clear Creek district breathlessly awaited the coming millennium. Dashed expectations arrived soon enough. The *Courier* of May 30 tersely noted that "it is currently reported that the Malm mill has been completely dismantled and all the machinery shipped away."

## PHOTO ESSAY 1

# GEORGETOWN AT THE TURN OF THE CENTURY

*Young cottonwoods in front of the Nelson home on Rose Street c. 1895. In later years the small brick building would serve as the Catholic rectory. Denver Public Library, Western History Collection, f5212.*

*These women referred to themselves as the first literary society in Colorado, which may be true, but difficult to prove, c. 1900. "Back row: Miss McKee, charter member; Miss Elizabeth Gaylor, charter member; Mrs. Henry Seifried; Mrs. Charles Wolfe. Middle row: Mrs. J. S. Randall, charter member; Mrs. Charles Bullock, charter member and first president; Mrs. E. S. Weaver, charter member; Mrs. Orson Clark; Mrs. Wm. Jacobs (aunt of Mrs. Pulsifer); Mrs. Dr. Collins, Sr., organizer; Mrs. Ella Adams. Front row: Miss Mattie Naylor, charter member; Mrs. Horace Bullock, charter member; Miss Ada Hardy, first secretary; Mrs. Frank Wood, first member to be admitted after the organization." Label courtesy of Erl Ellis collection (currently in possession of Christine Bradley). Colorado College Special Collections, Kernochan, pp. 82–22a, album 3 p. 56.*

*Women pose for a picture in front of Dalgleish photo studio, Sixth and Taos streets. Note the photo-drying racks on the second floor. Image courtesy of History Colorado (Dalgleish photo, f25107).*

*By the late 1890s Georgetown's commercial district covered every inch of available space, but paint was beginning to peel and roofs needed fixing. Image courtesy of History Colorado 10032956.*

*The parade winds its way up Taos Street c. 1900. The Dalgleish photo studio building is on the east side of the street. Stereoviews are visible in the windows, and a display kiosk is on the corner. Denver Public Library, Western History Collection, Dalgleish photo x1227.*

*The Kneisel and Anderson supply wagon stops for groceries in front of the building that still stands on Sixth Street. Denver Public Library, Western History Collection x6530.*

*The Wheeler family takes a family outing complete with bicycle, guitar, and cookstove, c. 1900. The family took many trips up toward Green Lake, but the exact location is unknown. Photo from the Wheeler collection, courtesy of Historic Georgetown, Inc.*

*Two young girls pose while looking over Georgetown, c. 1910. The cottonwood trees planted in the 1890s have leafed out, providing shade and a sense of permanence to the community. Image courtesy of History Colorado, f35085.*

*Georgetown's women don white dresses for their bicycle ride up Taos Street Monday, July 5, 1897. George Dalgleish Photo used courtesy of Historic Georgetown, Inc.*

*Members of the Georgetown Volunteer Fire Department dress up for a department photo in 1898. Image courtesy Georgetown Volunteer Fire Department.*

*The arc light at the intersection of Sixth and Rose Streets, c. 1905, marked a transition from the old gas street lights. Image courtesy of History Colorado, f15669.*

*The stereoview is labeled "Green Lake Avenue," a relatively wide thoroughfare for a mountain road. The wagon is headed to Green Lake, south of Georgetown, a popular tourist destination. Bradley collection.*

*The lots west of the Hotel De Paris, now used as a parking lot, were used as a garden for a short time after the removal of the foundations of the buildings burned along with the McClellan Opera House in 1892. Photo courtesy of Hotel De Paris, Colonial Dames of America, Colorado chapter, Burkholder collection.*

*The Burkholder children, James E. (Ned) and Hazel c. 1895, were raised at the Hotel De Paris in Georgetown. Photo courtesy of Hotel De Paris, Colonial Dames of America, Colorado chapter, Burkholder collection.*

*Taking the family on an outing, the adventurers pose with the rig before heading up Fourth Street to the Green Lake road c. 1910. Image courtesy of Historic Georgetown, Inc., Wheeler collection.*

*Wheeler family members head off for a pack trip to Green Lake c. 1910. The women are riding sidesaddle. Image courtesy Historic Georgetown, Inc., Wheeler collection.*

*The Wheeler family poses while on a picnic at Green Lake, c. 1900. Image courtesy of Historic Georgetown, Inc., Wheeler collection.*

*Mrs. Anderson holds son Henry, with daughter Marian, in front of their home at the corner of Second and Argentine Streets about 1910. Colorado College Special Collections, Kernochan, pp. 82–22a, album 3 p. 56.*

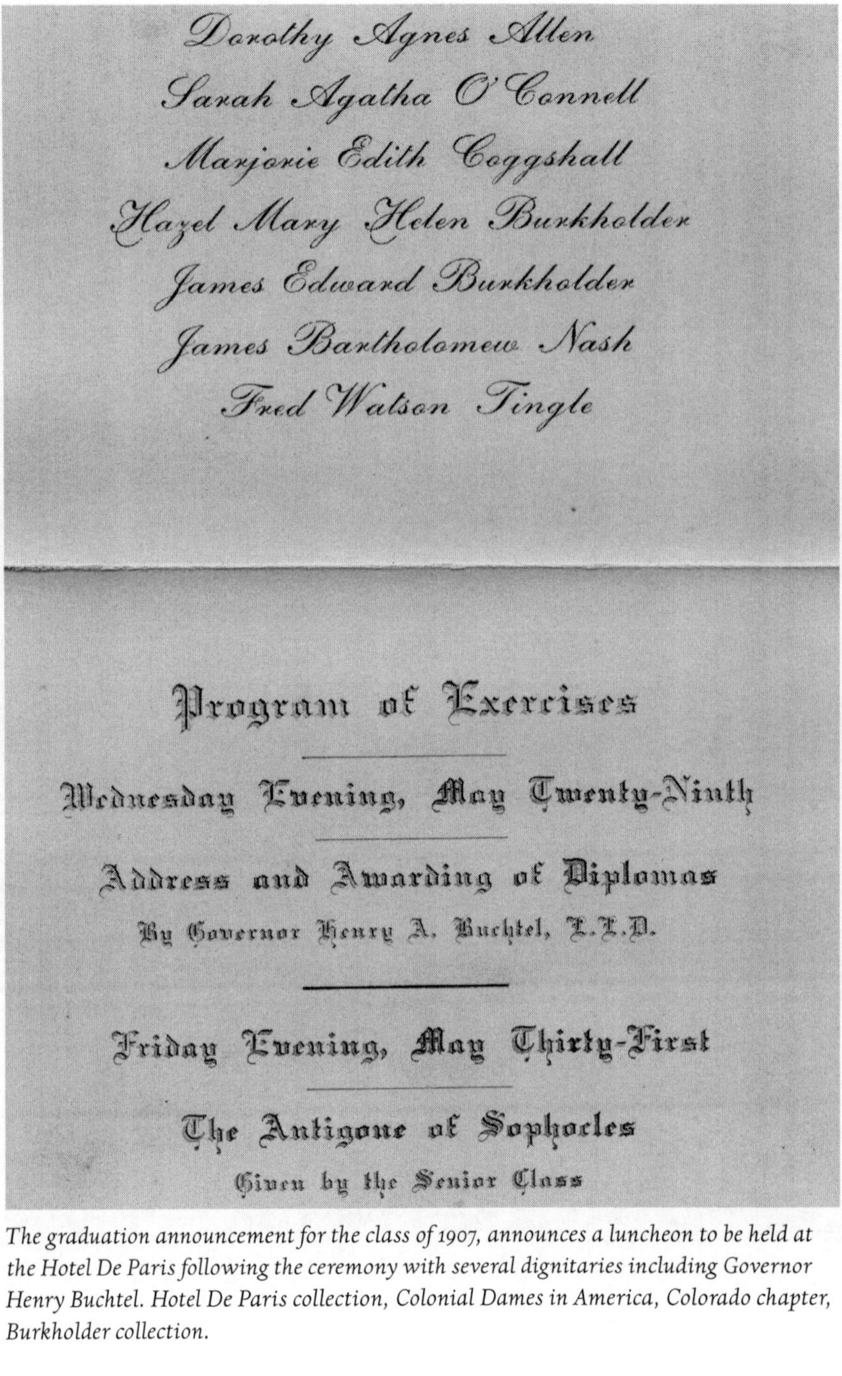
Dorothy Agnes Allen
Sarah Agatha O'Connell
Marjorie Edith Coggshall
Hazel Mary Helen Burkholder
James Edward Burkholder
James Bartholomew Nash
Fred Watson Tingle

Program of Exercises

Wednesday Evening, May Twenty-Ninth

Address and Awarding of Diplomas
By Governor Henry A. Buchtel, L.L.D.

Friday Evening, May Thirty-First

The Antigone of Sophocles
Given by the Senior Class

*The graduation announcement for the class of 1907, announces a luncheon to be held at the Hotel De Paris following the ceremony with several dignitaries including Governor Henry Buchtel. Hotel De Paris collection, Colonial Dames in America, Colorado chapter, Burkholder collection.*

*"Pop" Olds leads the Fourth of July parade striding along in front of the firemen's float c. 1900. The large sign behind him advertises Ladies Rest Rooms. Image courtesy of History Colorado, f8022.*

*There are times the number of people in a parade can outnumber the number watching, as shown as this Georgetown parade rounds corner of Sixth and Rose Streets, c. 1910. Photo courtesy of History Colorado, f24431.*

*Hazel and Ned Burkholder are seen with friends eating watermelon on the porch of the Hotel De Paris c. 1910. Note the protective covers some of the women are wearing. Photo courtesy of Hotel De Paris, Colonial Dames of America, Colorado chapter, Burkholder collection.*

*Ermina Forbes' dining room is set for company c. 1900. Image courtesy of Historic Georgetown, Inc., Wheeler collection.*

*Ruth Wheeler is ready to play the piano at the family home in Georgetown about 1910. Image courtesy of Historic Georgetown, Inc., Wheeler collection.*

## CHAPTER 2

# ALL DRESSED UP WITH NO PLACE TO GO

Colorado's mining communities bounced along with the winds of fortune, good and bad, as the nineteenth century drew to a close. The passage of the Sherman Silver Purchase Act by the federal government in 1890 spurred a building boom: new brick buildings in the commercial district and three beautiful, spacious residences (those of Thomas Cornish on Argentine Street, John Henry Bowman on Rose Street and B. F. Potter's extensive renovations to his recently purchased home on Fourth Street, later the home of Frank Maxwell). The good times were replaced with stunned silence in late June 1893, however, when Great Britain announced the cessation of silver coinage in India. The *Silver Standard* (Silver Plume) quietly announced: "Another blow to Silver; The Free Coinage of Silver in India Suspended and Bullion Drops to 77 Cents"[1] The action of a foreign government, thousands of miles from the silver mines of Georgetown and Silver Plume, shattered the local economy—closing mines, smelters, and the local bank. Three weeks after the announcement, the *Silver Standard* reported:

> A change as great as that ever pictured as being wrought by a magician's wand has come over this community in a few weeks time; a change that words alone can convey but a faint idea of. Then all was bustle and activity. Heavily laden ore wagons and jack trains were constantly busy carrying the product of the mines to the market, and loads of supplies were in turn sent to the mines. . . . All building or thought of it has been stopped; depression hovers in the air; groups of idle men congregate and quietly but earnestly talk over the situation and vainly wonder what the outcome will be. . . . Some have gone fishing; they might as well since there is nothing else to do. Besides they have a precedent. Even the president goes fishing in times like these.[2]

By August 1893 the Bank of Clear Creek County reopened its doors at the same time that President Grover Cleveland asked Congress to repeal the Sherman Silver Purchase Act, which he felt aggravated the instability of the national economy. The final blow to the future of silver mining came in November with the repeal of the act. The local economy staggered along with silver mining. While economic uncertainty had been part of town's history since the discovery of gold, in 1859, changes in the world economy along with harmful decisions on the part of the federal government created problems well beyond the troubles of the past. Some businesses left town, yet many decided to try to wait it out.

Three years later, the residents of Georgetown and Silver Plume still pondered "How long will it last?" and placed their hopes on the national election and the candidacy of William Jennings Bryan. The years between 1893 and 1896 had been troubled, yet the towns had regained their footings and continued to operate. The Bank of Clear Creek County secured the signatures of depositories to guarantee repayment of funds over eighteen months as follows: "The agreement is to the effect that depositors are to be paid 10 per cent of their deposits in 6 months; 15 per cent in 9 months; 25 per cent in 12 months; 25 per cent in 15 months; and 25 per cent in 18 months from the 15th of this month."[3] The optimistic voters of Silver Plume approved bonds for construction of a new schoolhouse in 1893 and a town water system the following year, major projects for a town facing economic uncertainty.

Merchants realized the importance of tourism—an industry without direct ties to the economic futures of the world metals market. The national depression of 1893 curtailed the number of visitors; however, tourists kept the town alive while the mines struggled—a trend that would continue well into the twentieth century.

The railroad, known as the "Far-Famed Georgetown Loop" and the first-class accommodations at Louis Dupuy's Hotel De Paris and at numerous smaller hotels and guest houses attracted visitors from around the world. Pack trips to Green Lake or the longer trip to Gray's and Torrey's peaks continued through the hard times. Mountains and cool summer days beckoned unabashed. The railroad hired photographers and published brochures aimed at the traveling public.

Even with an influx of tourists, however, the population dwindled through the 1890s, and many of the town's professionals—attorneys, dentists, physicians—left in search of clientele. Those remaining in town, however, continued with the activities and organizations of a typical nineteenth-century community. Fraternal and social organizations remained prominent. In 1897 the *Silver Standard* reported, "There are sixteen secret societies in Georgetown, and about the same number of social clubs."[4] The Methodist, Catholic, Episcopal and Presbyterian churches stayed active, along with the newer Swedish Lutheran church, which now met in the former Congregational Church building on Fourth Street. The Catholic and Lutheran congregations expanded as immigrants continued to arrive from overseas, especially Swedish and Italian immigrants anxious to find work in the mines.

The future of the industry was in serious doubt; nevertheless, Clear Creek County mines continued to produce ore. Leasing of individual drifts increased in an effort to allow a man to work his own claim and be rewarded with any possible profit instead of "day labor"—working for wages that had dropped at this time from $3.00 per day to $1.50. The *Silver Standard* explained: "The leasing system of working mines seems to be gaining in popularity in all all [*sic*] parts of the state where mining is carried on. Clear Creek and Gilpin counties were the first to adopt the system, but it has been gradually extended until there is hardly a camp in the state where it is not in vogue. It offers a chance to the miner to make something more than day's pay, while it is more economical to the mine owner, as he gets his mine opened up with the expenditure of less money than where he hires everything done."[5] Decades of detrimental legal wrangling decreased as remaining companies learned to consolidate surrounding claims, often owning whole mountainsides in an effort to thwart the maverick mine owners who had ground operations to a halt with lengthy legal battles in the past.

Efforts turned to reducing the costs of producing ore—new tunnels, new mills and new technology. Chief among the technological advances was the introduction of electric power, more specifically the availability of alternating current to transport high voltage through three-phase wires directly to the mines, many of which were perched on hillsides or located in remote valleys miles from previous sources of power.

Georgetown's two creeks—South Clear Creek and the South Fork of South Clear Creek, names that have created over a century of confusion—provided power from the earliest days. Creek waters turned waterwheels, which powered rudimentary mills. In 1891 the Electric Light Company, under the leadership of George W. Hall, acquired the water rights of the Farwell Mining Company along South Clear Creek and commenced construction of a small power plant near the Hall lumber mill at the west end of Fourth street, using the reservoir and penstock running from the area known as Devil's Gate located approximately fifty feet downstream from the high bridge of the Georgetown Loop Railroad. Plans were made to light all of Georgetown and Silver Plume. The plant went into operation in June 1891.[6] The *Courier* proudly announced: "The present plant can furnish 700 lights and 60 horse power for hoisting during the day."[7]

The young company faced problems. The townspeople seemed to favor electric power but expected consistency from an industry that depended upon the available flow of water in a mountain creek. A prior agreement with the Baldwin Mining Company for the water rights of the Farwell Company included a requirement for electric power, except for October 15 through April 15, when it was assumed the available flow would not be sufficient to power the mining operations. Homeowners were unwilling to go six months without power. Ice also posed a problem during the winter months. In December 1891, "Ice choked up the electric light company's water pipes last Sunday night and left the town in darkness for an hour."[8]

Locals worried that the company would not be able to compete in the same market as the gas company. The town's first electric company had shut down in 1887, unable to secure enough of the local market.

Fears of explosion and fire further complicated matters, especially as inventor Thomas Edison spread rumors about the danger of alternating current (undoubtedly fueled by his desire to have the industry use his patents related to direct current). By the time the new Georgetown plant opened in 1891 the technology relating to lighting homes, especially through the use of water power, was relatively safe, if inconsistent, yet the potential for mining and heavier industrial uses was just starting to develop.

In July 1891 both Denver and Georgetown papers reported the successful transmission of power, through the use of alternating current, from the far-off Ames hydroelectric power plant to the Gold King Mine outside of Telluride, a distance of over two miles. The genius of Nikola Tesla, combined with the business acumen of George Westinghouse and Lucien Nunn, provided the ability to transmit power over distances not possible with Edison's direct current systems.

For the mining industry, the potential was limitless:

> In nearly every camp there is an abundance of water power to be had, but the trouble is to get the power just where it is most needed, or the ore to where the power could be utilized. This problem seems near to solution by the combination of the water power and the electric motor and the practical methods for the transmission of the power so generated directly to the mine.
>
> Recent experiments in the utilization of electricity in this manner have been watched with a great deal of interest by miners and mine owners, and the success of the one at Telluride is of no small importance.[9]

Almost immediately the young Georgetown company announced plans to expand its operation, adding machines capable of producing alternating current to provide power to the mines and mills in the area. In 1893 Georgetown and Silver Plume would see the potential develop with the formation of the United Power and Light Company of Clear Creek County. The April 14 edition of the *Georgetown Courier* reported on a meeting at the offices of the Bank of Clear Creek County: "The first installment of 50% of the subscriptions to more than half of the capital stock was paid in, which insures the success of the enterprise, as it is understood that the Bank of Clear Creek County will take the balance of the stock that is not subscribed."[10] The list of incorporators included four members from Georgetown (Fred P. Dewey, Robert O. Old, Fred Pneuman, and Ensign S. Wright) and three members from Silver Plume (Casper Desch, Phil L. Roberts, and J. M.S. Egan). The same issue noted that "the Silver Plume people were out in force and were enthusiastic over the prospect of soon having light in the prosperous town." The young company acquired the assets of G. W. Hall's Electric and Light Company, specifically the small plant near Hall's Lumber Mill.

By June 1893 the new United Light and Power company had completed electric lines to Silver Plume. An upgraded pipeline, an additional dynamo, and a new Pelton water wheel made the transmission of power to Silver Plume a reality. In all likelihood, the new dynamo provided alternating current in order to be able to cover the two-mile distance between the two towns.

In order to acquire a franchise from the town of Georgetown for transmission of power, the new company bought out the Georgetown Gas Company. The *Courier* noted: "The result will undoubtedly be an increase in the rates."[11]

Electric power quickly became the standard in both communities, and the gasworks closed in 1894.[12] The town fathers had ordered the gas lights turned off in August 1893 in response to the economic panic.[13] The merchants and residents complained, but the town was unable to pay its bills. G. W. Hall and friends installed high-powered arc lights—one at the corner of Sixth and Rose and the other on Chimney Rock: "It throws rays to every part of town."[14]

Two years later the town fathers felt solvent enough to light the streets once again: "At the request of a number of citizens, Mr. Arnold brought up the subject of lighting the streets. He believed that 50 lamps would be sufficient, and ascertained from the Electric Light company that the company would furnish and light that number at the rate of $1.50 per month for each lamp. He also ascertained that Idaho Springs is paying $2.45 per month for like service, Central City $2.66, Golden $1.25, and Silver Plume $2. Mr. Arnold believed the town could afford the expense, as it had succeeded in wiping out a large debt last year."[15] The town authorized a proposal for fifty-five lights at $1.20 each per month, with the company to furnish the globes.

In 1899 the company announced that it would be installing a three-phase generator to provide power to the Maxwell Mill at the Mendota Mine in Silver Plume.[16] Nine thousand pounds of wire and a new thirty-five horsepower, three-phase generator were delivered to Georgetown.[17]

As the use of electricity increased, the need for a steady flow of water for the hydro-based system soon overwhelmed the available resources of South Clear Creek. Attention turned to the "left hand fork"—the South Branch of South Clear Creek. While the volume of water was not as great, the potential for a system of reservoirs to supplement the flow during periods of low water predicated a move to the other side of town. In addition, the potential of meeting the energy needs of the gold mines around Freeland, Lamartine, Idaho Springs, and Central City provided the impetus to build a new plant with much greater capacity.

In January 1900 United Light and Power acquired the old sampling mill at the east end of Silver Plume to be torn down and used in the construction of a new hydro-plant at the east end of Sixth Street.[18] The Georgetown property—which had housed the early Mount Alpine Mill, the What Cheer Mill, the German Reduction works, and the Chamberlain Sampling Works—would continue as a lynchpin in the mining community of Georgetown. Negotiations turned to the acquisition of the Clear Creek Company works further up the creek at the east end of Fourth Street. The power company moved to acquire the property as well as the water rights and pipeline. The town council entered negotiations relating to the pipeline, since the town relied upon the line for fire pressure.[19] The company's water rights and developing chain of reservoirs put it in a position of strength compared to a number of smaller power companies beginning to appear along the Clear Creek Valley.

The Clear Creek Power Company, under the direction of A. R. Specht, a banker out of Boston, had acquired property and storage rights in the vicinity of Dumont.[20] In addition, the company acquired rights for a storage reservoir above Empire, which the local papers thought would have a secondary recreational benefit.[21]

United Light and Power moved quickly to secure its position and the contracts with mining companies that were needed to finance improvements. Construction of the new plant at the east end of Sixth Street started early in 1900. At the same time, the company began laying 3,500 feet of pipeline to a small dam (a short distance above today's town water storage tank), approximately 480 feet above the intake at the new plant. The local papers extolled the operation: "We feel certain that Georgetown will feel the life of the new improvement, and we certainly wish our citizens who are behind this enterprise all the success which their public spirit and their large expenditure entitle them."[22] The capital expended on the plant and its infrastructure provided work for many men in town who had been struggling to find steady employment since the hard times of 1893.

The new plant powered up in October 1900, though the penstock and dam were still under construction.[23] The equipment housed in the new building overwhelmed the smaller operations popping up around the county:

> It will be the most powerful electric plant for power purposes in the state outside of the Cripple Creek district. The building has been made large enough to double the capacity if necessary. To start, there will be two water wheels of Denver manufacture, two powerful Babcock and Wilcox water tube boilers, one Corliss engine, and two new electric power generators. For eight or nine months in the year, there will be ample water to operate the plant. During the low water months steam power will be used in connection with the water power. The pipe line is probably the best in the state, steel pipe nearly one-fourth of an inch is thickness being used. The pressure will be 212 pounds to the square inch.[24]

Plans for plant expansion began almost as soon as the new facility opened its doors. The company signed an agreement with the Cascade Electric Company of Idaho Springs whereby Cascade agreed to run lines to connect with the United Light and Power lines in Lamartine Mining District at the head of Trail Creek. Cascade Electric and its president, Lafayette Hanchett, had already secured a franchise from the city of Idaho Springs.[25] Hanchett's investments in Lamartine mining property put him in a position to negotiate an agreement beneficial to both sides.

In 1902 the competition for business heated up when William Renshaw's Seaton Mountain Power company acquired the Judd and Crosby water rights and Farwell Mill properties for $5,500 with the intention of piping water through town and constructing a hydro-plant just north of the old Wilson and Cass millsite at the north end of town.[26] Renshaw believed that the original power plant site along South Clear Creek could be redeveloped and improved, though townspeople felt that redirecting and piping the creek presented several sanitary issues.

United Light and Power bought additional water and reservoir rights on the South Branch of South Clear Creek.[27] The company soon began constructing a new dam, just north of Silver Dale: 20 feet high, 24 feet across at the base and 14 feet across the top. The resulting reservoir and 2,000 additional feet of penstock were expected to triple the capabilities of the company.[28] The *Courier* noted, "The dam is a most substantial affair, and is being built for all time."[29]

At the same time, the *Idaho Springs Gazette* noted that control of the company had transferred to Idaho Springs parties.[30] Management of the company remained the same, but Lafayette Hanchett of Idaho Springs and Fred Himrod of New York, major investors in the Cascade Power Company, bought out the interest of the Clear Creek Bank, thereby allowing the bank to pay off the rest of its debts related to the panic and its closure in 1893. The new board moved quickly to acquire the properties of the Green and Clear Lakes Company and series of lakes and reservoirs above its new reservoir. Dams were constructed at both Clear and Green Lakes in 1902, increasing the amount of reservoir storage and allowing the company to have better control of the water flow of the entire drainage. The company added twenty feet to the new Georgetown plant to allow for additional equipment and output. Agreements were reached to provide electricity to Central City, Nevadaville, and Black Hawk. While the good people of Georgetown took pride in their local hydroelectric industry, the nature of the investment was changing. First Idaho Springs, and then other regional and statewide interests would crowd the market and create a conglomerate as unwieldy and tempestuous as the railroad industry.

Colorado's mining communities knew that they couldn't survive without a railroad and also knew that these same railroads compounded many of their economic woes. Mine owners needed to ship large quantities of ore to smelters in Denver or Leadville. Railroads struggled to survive with the hard economic times, and several smaller lines went bankrupt. Commercial and mining interests bore the brunt of the hard times. Populist platforms in the statewide election of 1893 and the national elections of 1896 and 1900 included the need to regulate the railroads along with discussion of the possibility of government ownership.

In the midst of all of this turmoil, two investors decided to try to develop a railroad from Silver Plume to the newly opened (and a few reopened) mines in East Argentine. Once again, the area that had produced enough high-grade silver ore to start the major rush to the area in 1864 stepped forward with a combination of new and old mines with a valuable mixture of gold and silver ore. In 1899 William Rogers prospected the area and staked his claim to the Santiago Mine. Shortly thereafter, Edward J. Wilcox discovered the Waldorf and Argentine properties. A small community first known as Wilcox, later Waldorf, sprang up near the mouth of the Argentine Tunnel. In an age of economic uncertainty, Rogers and Wilcox both wanted to develop a railroad line that would have been questionable in the best of times.

William Rogers's Hazelton Company was the first to begin construction. The company owned and operated the Baltimore Tunnel and associated properties about a half mile west of Brown Gulch in Silver Plume. Rogers saw a benefit to connecting this property with his Santiago properties on McClellan Mountain. In 1904 he attended a meeting of Silver Plume's Board of Mines and Trades (a group of mine owners a businessmen formed to publicize and support local interests) and announced plans to lay ten miles of track from Silver Plume to East Argentine and to build a new mill at the mouth of the Baltimore Tunnel. Rogers asked the group "what inducements the citizens of this town would offer toward the new enterprise."[31] The reactions were positive to say the least. The *Silver Standard* promised an economic boon: "What this railroad means for Silver Plume can hardly be imagined. We would soon have a population of 5,000 people, mills, banks and several other business houses and also one of the liveliest mining camps in Colorado."[32]

Within one week the Board of Trades agreed to provide men and equipment to build the first mile of track. Rogers traveled east to work on more funding for the enterprise and his company, now known as the Hazelton-Santiago. The railroad venture incorporated in 1904 as the "Gray's Peak Railroad Company" with 2,500 shares available for $100 each. The company hoped to build a railroad from Silver Plume around the face of Leavenworth Mountain to the southeast base of Gray's Peak, then over to Keystone, and later to Minturn and Wolcott in Eagle County.[33] Work commenced in June 1904 with twenty men, possibly the volunteers promised by the townspeople. In August Gabriel Bartolomea received a contract for 1,000 feet of grading starting across from the Colorado and Southern pavilion, while the employees of the Hazelton-Santiago Company started to build the road west to connect with the properties around the Baltimore Mine. With considerable understatement, the paper reported: "There are some large ledges of rock in that vicinity through which a roadway will have to be blasted."[34]

The "large ledges of rock" outlived the project, and while at least one mile of Rogers's dream was built in summer and fall 1904, the roadway could never be used since the blasting was never completed. Rogers's project slipped into history after one year of work.

Meanwhile, Edward Wilcox shared a similar vision and moved to start construction toward his properties in 1905. The Argentine Central Railway Company incorporated in 1905, offering 500 shares of stock at $100, for construction of track from Silver Plume to the summit of Mt. McClellan.[35] Work commenced August 1, 1905, with a new (and more practical) route out of Silver Plume that skirted the substantial cliffs southwest of town. By November, the tracks rounded the point of Leavenworth Mountain, and the papers commented on the novelty of seeing trains on the track above town.[36] The Wilcox dream came to fruition in January of 1906, with completion of the road to the Wilcox Tunnel, and a golden spike driven as part of an "impressive" ceremony where "no great pomp was observed." After the spike was driven, Superintendent C. L. Tingle broke open a bottle of "aqua pura," undoubtedly in deference to Mr. Wilcox's temperance proclivities, christening the new town "Wilcox," a name that would shortly be changed to "Waldorf."[37] The high hopes and heavy investment provided Wilcox with the ability to ship ore directly from his high altitude mines as well as providing a thrilling ride to future tourists.

Completion of the railroad meant that William Rogers and the Hazelton-Santiago properties could also have access to rail transport, yet the two men never came to an agreement. In 1905 the Hazelton-Santiago Company purchased the Dewey Mill on the old Clear Creek Mining Company's property at the head of Fourth Street in Georgetown. Rogers told the local papers that the acquisition of an existing mill would open the operation immediately instead of spending months building at a new location (not to mention the money he would have to pay to Wilcox for transportation of materials to the high country). Instead, ore from the Baltimore Tunnel property could be easily transported to Georgetown.[38]

The small town of Waldorf (nee Wilcox) struggled through snowstorms and avalanches, producing ore but also producing headaches. A third group of investors—known as the East Argentine Tunnel, Mining, Milling, Power, and Transportation Company—had came on the scene in 1902. Composed entirely of local businessmen, the company proposed construction of a tunnel in Waldorf well below the other existing operations, as well as an additional tunnel and power operation at the area between Silver Dale and Waldorf soon to be known as Sidneyville. The Board of Directors came straight from the local business directory: "Marcus Wheeler, president;

Henry Kneisel, vice-president; R. A. Leisy, secretary-treasurer; M. Sidney, manager; J. J. White, counsel; J. S. Randall."[39]

Boardinghouses sprang into existence as everyone tried to get under cover before the snows overwhelmed the enterprise. Hearty souls worked through the winter months in a effort to extract enough ore to pay for wages and investments as well as transportation of the ore to mills and smelters. The lack of trees made the hillsides treacherous and prone to snowslides. In March 1906 the workers at the Santiago Mine awoke with a jolt:

Before day break on Friday morning of last week, a snowslide swept down on the buildings at the Santiago Mine, imperiling the lives of Mrs. Tatro, Si Newman and seven Italian miners. The buildings, consisting of the boarding and bunk houses, blacksmith shop and stable, were reduced to kindling wood. The inmates were asleep, and their first realization of what had taken place was on finding themselves sprawling in the snow, with the exception of Mrs. Tatro, who was still lying upon her bed which had been carried a short distance down the mountain. Amid all the wreckage of crushed buildings, not one of the inmates received so much as a scratch. After gathering their wits and finding there was no clothing to be found, the party started barefooted and with scant clothing for the Paymaster Mine, nearly half a mile distant, which they reached in safety but with badly frozen feet and hands.[40]

Undoubtedly the Italian miners flashed back to the deadly snowslide above Silver Plume in February 1899 in which ten Italians (mostly women and children) had been killed in Cherokee Gulch. Fortunately, no one was killed or badly injured in the Argentine slide.

In 1904, prior to the start of railroad construction, Wilcox had contracted with United Light and Power to extend electric power to his mines in the vicinity of Waldorf, while also starting plans to tunnel west underneath McClellan Mountain to connect with his Stevens Mine property at the head of Quail Creek (Stevens Gulch). The cost of running an electric line to the mine must have been prohibitive as was the cost of drilling a tunnel through a mountain. These expenses, however, included hundreds if not thousands of hours of labor—work for local miners and others struggling to survive in an economy still hoping to revive the "days of yore." In many ways the economic revival the papers were hoping for after the national depression of 1893 came with the ill-advised investments of mine owners trying to save their foolhardy plans with even more money, all the time looking for the right combination of location, technology and good luck—generally to no avail.

Perhaps the only solid investment came from the owner of the Mendota Mine in Silver Plume who began giving tours of his property to tourists in 1904: "Electrick [*sic*] lights have been installed at the Victoria tunnel and the management is now ready for the summer tourists. Thousands of people from all parts of the globe are shown through the Mendota Mine every season, and we believe it a splendid way of advertising this little camp."[41]

PHOTO ESSAY 2

# THE PHOTOGRAPHY OF LEWIS CLARK COGGSHALL

By the turn of the century, the Colorado and Southern no longer hired professional photographers to travel to Georgetown and Silver Plume for beautiful portraits of the mountain towns and the high bridge of the Georgetown Loop. William Henry Jackson was off to other locations. George Dalgleish had a storefront and photo gallery at the corner of Sixth and Taos streets, where he produced stereoviews, "cartes de visites," of local individuals, groups, and scenery. Fortunately, Lewis Clark Coggshall moved to Georgetown to work for the power company about that time. The young man brought along his professional-quality camera and started taking photos of his friends, family, scenery, and occasional photos commissioned by local companies. With an artistic eye and a love of the mountains, the Coggshall photos give us a wonderful insight into the lifestyle of turn-of-the-century Georgetown and the surrounding areas.

The Coggshall family kindly shared copies of individual photos and a scrapbook for this publication.

*Lewis Coggshall, amateur photographer, stands by the Devil's Gate bridge above Georgetown c. 1915. Coggshall family photo.*

*The family lived in the lovely old home at 921 Rose Street. Coggshall family photo.*

*Horses and the camera setup made their way to the top of Gray's Peak, altitude 14,270 feet. Neal Randall, Tom Hamill, and L. C. Coggshall are labeled in this photo from one of the family scrapbooks. Coggshall family collection.*

*Coggshall had his camera along on this trip to the West Argentine Mining District up Stevens Gulch. Coggshall family photo.*

*Coggshall's artistic skills are exhibited in this photo of lightning hitting the saddle between Gray's and Torrey's peaks. Coggshall family photo.*

*Women in summer dresses and men in ties enjoy a summer picnic. Coggshall family photo.*

*The women are ready to go underground as part of a mine tour, probably taken at the entrance of the Burleigh Mine on the west end of Silver Plume, with the Argentine Central Railroad grade in the background. Coggshall family photo.*

*This road (possibly west of Silver Plume) made for a nice automobile drive. Coggshall family photo.*

*Old meets new on Berthoud Pass, as cars and donkeys try to figure out how to pass. Coggshall family photo.*

*The road west of Silver Plume, just east of Graymont or Bakerville, winds through trees and bushes now destroyed with the construction of Interstate 70. Coggshall family photo.*

*The gentleman and young man in the middle don't look very excited as they all pose in front of the old Jennings house in Graymont. Coggshall family photo.*

*The photo is labeled "On way to West Argentine, Big Professor." Big Professor would be renamed Mount Sniktau in 1926. While automobiles were starting to drive up some of the old wagon roads, some roads remained too steep or rocky for cars. Coggshall family photo.*

*"Snowball fight on way to Gray's Peak—Irene Coggshall, Marjorie Coggshall & Mrs. C." Marjorie Coggshall, sister of photographer Lewis, would marry Conrad Criley in 1912. Coggshall family photo.*

*The ladies may have just crossed a small creek in search of wildflowers. The photo was probably taken in the West Argentine area near the trail to Gray's Peak. Coggshall family photo.*

*Photographer Lewis Coggshall poses in the door of a miner's cabin, possibly on Saxon Mountain. Coggshall family collection.*

*Skiing started to gain some popularity in the early twentieth century in Clear Creek. Photographer Coggshall was ready to capture the adventure. Coggshall family photo.*

*Rocks appear to block the road for all but horse or foot traffic, while everybody takes a rest along the side of Green Lake above Georgetown. Coggshall family photo.*

*This good-sized cabin may have been associated with the Criley's High Five Mine above Silver Dale. Coggshall family photo.*

*The men in this rare nighttime photo may have been sharing a few tall tales at the Naylor Lake Fishing Club. The label says it is "Outside big cabin at Naylor." Coggshall family photo.*

## CHAPTER 3

# THE DOWAGER SILVER QUEEN

> A fire involving a loss of several thousand dollars broke out about half-past ten o'clock last Saturday night in the rear of some small vacant store buildings on Sixth street, between Rose and Argentine, and almost completely destroyed these buildings and damaged several others to a greater or lesser extent.
>
> Georgetown Courier, *July 11, 1908*

The townspeople of Georgetown maintained an optimistic outlook even after the economic upheavals of the 1890s. Many residents left, yet there seemed to be enough work for those who remained. Work at the mines was sporadic, but construction of new mills, railroads, and the new power plant and associated power lines, as well as occasional road projects, kept food on the table for most families; however, the town was aging. Younger families and professional men began to search for opportunities in new, up-and-coming parts of the country.

The Fourth of July fire left another gaping hole along Sixth Street, in the town's commercial center and in the hearts of the townspeople. Sparks from fireworks fell onto aged, parched roof shingles, a sad twist of fate for a town that enjoyed this holiday above all others. Memories of hose cart races and parades through the years were now replaced with the smoldering ruins of former businesses. The fire consumed buildings in the center of the business district in the vicinity of today's library building, two blocks away from the former site of the McClellan Opera House, which had burned to the ground on January 11, 1892, and was never rebuilt.

Several smaller fires obliterated other parts of the downtown: the Barton House annex was consumed December 14, 1901, and the Hall Mill and Lumberyard at the west end of Fourth Street near the base of Chimney Rock on April 17, 1902. The Santiago Mill, formerly known as the Taylor or Clear Creek Company Mill, at the east end of Fourth street would succumb to fire on August 26, 1910. Commercial buildings on the east side of Rose street between Fifth and Sixth streets went up in smoke in July 1918. Various companies of the volunteer fire department and their ever dwindling number of members continued to respond, saving the rest of the town's aging buildings.

The four companies of the fire department continued to sponsor tournaments, including "the largest money prizes ever given in the state" in 1903. Prizes included "Hook and ladder race $170 first, and $50 second. Plug race $125 and $50. Straightaway race $125 and $50." After 1905, interest in the fire department waned: "All interests seems to have dropped in the fire department, the department holding only their annual meetings, and most of the companies, none at all after 1909. What caused this lack of interest on the part of the citizens in so important an organization is hard to discover. But such apathy as this appears once in a while in nearly all things."[1]

The four individual companies struggled to keep the minimum of fifteen members each. A new wave of interest would come in 1916 with the election of new officers and a new, younger generation of men willing to sign up and to protect the town and show continued interest in the department's history.

While the mining industry struggled, unemployed miners in Georgetown and Silver Plume still found work, even if the industry was dying. A flurry of new mills provided employment: building the mill, operating it (although some never opened their doors), and tearing it down to move and rebuild elsewhere. In 1905 William Rogers, of the Santiago Mine, abandoned plans to build a mill near Waldorf in the Argentine District, instead purchasing the Dewey & Wheeler Mill (the old Clear Creek Mill at the head of Fourth Street) in Georgetown, which would only require "a few changes in machinery."[2] The next year, a fifty-ton concentrating mill was built near the mouth of the Kelly Tunnel northwest of town to process ore from the mines on Democrat Mountain.[3] At the same time, the owners of the East Griffith property across the valley started construction of another new mill in hopes of reviving the dreams of gold first envisioned by the Griffith brothers.[4] The gold reserves of the Capital Prize started to show promise at the same time, and the owners soon decided to build yet another mill between Main and Griffith Streets in 1907.[5]

The good folks of neighboring Silver Plume were not to be left behind in this frenzy of mill building. In 1902 the Elkins mining company started construction of a new 100-ton mill at the Terrible.[6] In 1904 a new mill was under construction for the Seven Thirty properties, located just west of the Burleigh Tunnel.[7] In 1906 the buildings at the mouth of the Diamond Tunnel were torn down to allow for construction of the new Nashotah Mill the following year. As expected, "this mill is said by experts to be the most modern and best equipped to be found in Clear Creek county, the capacity being 150 tons daily."[8] At the same time the Scotia Company also built a new fifty- to sixty-ton mill on the east side of Charles Street between the creek and the railroad station.[9] In 1907 Messrs. A. W. Bradley, G. S. Love, and A. B. Loud started construction of yet another mill for the Jewell Mining Company, which was operating the Mendota Mine at the west end of town.[10]

The most interesting promise of "new and improved" mill technology came with the name of Thomas A. Edison. Metallurgical experiments in the late nineteenth century had not proved successful for the Wizard of Menlo Park; however, the owners of some of Silver Plume's old producers turned to Edison's latest technology to provide the answer. Suddenly Edison's name appeared in every issue of the local paper. Plans for the construction of an Edison Mill superseded all other mining notes.

> It is claimed by those who are on the inside that some of the very wealthy men of New York are back of this undertaking, and that Mr. Edison himself is personally interested. It is not a stock-jobbing scheme as there is no stock for sale, the promoters being willing to devote their own money to demonstrating the Edison process on a large scale, which if successful, will be of inestimable benefit to the mining world, as it is claimed a profit can be made on ore that carries as low as $3 a ton in metal values.[11]

A few weeks later:

> Thomas A. Edison, the electric wizard, who has been experimenting on ores from this region for the past two years, has concluded to erect his first exper-

imental works to prove out his method on a large scale, at Silver Plume, and will utilize the ores from the Dives-Seven-Thirty Mines.[12]

A letter from Edison to a prospective investor changed this perspective:

My Dear Sir: Your favor of the first instant has been received and its contents noted with much interest. In reply I would say that Henry B. Clifford's statements to you are correct as to the grinding and concentrating process, except as to cost per ton. I have no data on this and Mr. Clifford's figures seem to be low. However, I can say that we have operating the fine crushing rolls for over seven years at our cement works at New Village, N. J., where you can see them in operation. Clifford is an extreme optimist and I keep advising him to go slow and build a small mill at first and "try it on the dog," so to speak. If you investigate personally, I think you will find that the process we have here is out of the regular line and quite different from the schemes of the usual process man.

*Yours very truly,*

*THOS. A. EDISON*[13]

When J. B. Ballantine, one of Edison's engineers, arrived in town, the mining community felt the answer was at hand; however, Edison's chemical electric process didn't fare any better than the process tried at the Malm Mill in Georgetown. Construction would stop and start while negotiations for ore also slowed progress. Three years later, the mill was ready to start. After a couple of months, however, Mr. Ballantine modified the mill to the more mundane flotation process, and discussion of the Edison process dropped from the pages of the local paper.[14]

A short-lived boom in mining production came in 1916, spurred by a rise in the value of silver and by-products such as zinc and lead, as the war in Europe created a strong market. Silver Plume's mines and mills sprang to life: "Several hundred men are at work in the mines and what insures the great growing prosperity of the place is that nearly all the men are actually engaged in producing pay ore. This makes it necessary for many to be employed in the mills, all of which are running at full capacity. Every available team and wagon is at work hauling the output of the mines and there is an ever increasing demand for more teams and for mining timbers, the want of the latter being badly felt."[15] George Tarkington had started giving tours of the Burleigh Tunnel in 1915.[16] Perhaps the visitors were able to see mining firsthand.

Tourists continued to travel to the mountains in support of the one industry that remained strong. The railroad and stage lines brought visitors to the high country, but a new mode of transportation would soon change the future of Colorado. The production of automobiles, especially the Ford Motor Company's Model T and Model A, would soon bring travelers to the high country on their own. S. A. Ionides, chemist for the ill-fated Malm Mill, was one of the first residents to have his own car, a Cadillac, which arrived in Georgetown in February of 1910: "S. A. Ionides drove up from Denver Monday in his new 60-horse power Cadallac [*sic*] touring car. The machine is not built for racing but is prepared to make a steep grade with the greatest ease. A number of Georgetown people were favored with a ride in the automobile which everyone concedes is a beauty."[17]

The network of wagon roads throughout the county began to change as automobile traffic increased. Many of the early wagon roads were only eight- to ten-feet wide, with occasional pullouts, as loads had usually been taken uphill in the morning and downhill in the afternoons. Horse-drawn wagons could travel up steeper routes than autos. While the earliest automobile drivers were generally willing to back up the steepest sections of road (using the weight of the engine to help with traction and keep the gas flowing to the engine), that would not remain a reasonable solution as the number of automobiles on the roads continued to increase. The transition from horse to horsepower was not without humor, however: "Charley Nelson, who recently invested in a Ford machine, has become quite an expert driver, but when he does get stuck on a steep hill, strong Scandinavian language hasn't the effect on a Ford that it has on a span of Nelson and Pearson's thoroughbreds."[18]

The cost of maintaining mountain roads had overwhelmed the county and towns throughout the years. Many of these early routes had been developed as toll roads in an effort to provide some funds for maintenance. By the 1890s, however, the roads were in rough shape and the toll road companies did not have the funds to provide the necessary maintenance. Clear Creek County began to reclaim these rights of way, either through purchase or, when possible, legal maneuvering. In 1890 the county purchased the old Watson wagon road, the "Georgetown and Breckenridge Wagon Road," west of Silver

Plume. The *Courier* noted, "with May Day the toll gathered on that portion of the road will retire into appreciated absence."[19] In 1891, "the road from Georgetown to Hot Sulphur Springs [Berthoud Pass] is now free from any toll. The sum of $15,000 was paid for the Middle Park road to Junction ranch and $5,000 for the Rollins road from the Junction to Hot Sulphur Springs."[20]

In September 1893 the Board of County Commissioners declared the road between Georgetown and Silver Plume to be a free road. The charter of the toll road company had expired and, even though the incorporators were filing renewal papers while the commissioners were meeting, the county's legal representatives held that once the charter expired, the road was free and could be claimed by the commissioners.[21] The road from Silver Plume to the Seven Thirty Mine was declared a public highway in 1899, marking the end of toll roads within the county.[22]

State and federal highway dollars would become more important to the mountain counties in the twentieth century. Major highway projects, such as improvements over Floyd Hill or Berthoud Pass, required extensive resources. State and federal agencies realized the need for reliable routes and well-maintained roads and began to join in partnerships with local governments, especially since automobile owners began to apply political pressure. The national "Good Roads" movement started in the early 1900s. The Colorado Auto Club, established in 1902 along with the American Automobile Association, soon joined forces with the Colorado chapter of the National Good Roads Association established in 1905, and began lobbying the state legislature to identify and improve roads across the state. In 1909, the Colorado Highway Commission was created by the legislature with big plans and a tiny budget. Designated routes, many of which would soon have nicknames, were encouraged by local chambers of commerce in an effort to attract travelers. Circle routes gained popularity as families began to explore Colorado's mountains on day trips.

Reese Vidler, one of Georgetown's successful investors (who married into a wealthy family willing to help with some of his projects), formed a partnership with British interests—the Lookout Mountain Resort Company—to purchase land on top of Lookout Mountain, west of Golden. They hoped to develop a tourist road from Denver, through Golden, to Idaho Springs and Georgetown, continuing up to the Continental Divide then through the Vidler tunnel into Summit County. While considerable funds were invested and a funicular railroad built to the top of Lookout Mountain, plans for the road fell to the wayside.

Denver's Chamber of Commerce and an assortment of visionaries soon began exploring the mountains west of Denver in an effort to develop access to scenic areas while protecting view-scapes and forestlands. Discussions in 1910 soon moved to the creation of a Parks Commission, the hiring of noted landscape architect Frederick Law Olmstead Jr., a mill levy to pay for acquisition of lands, and the construction of "quality" scenic roads. After arguing with Jefferson County over the need for an eighteen-foot road—the commissioners thought the traditional eight-to-ten width for a wagon road would be sufficient—the Lookout Mountain road out of Golden started a new era in road building in 1913. Roads began to wind around mountain peaks in search of scenery and camping instead of just running directly to mines and mills. Within the next few years, the ever-growing number of circles designed for automobile travel made their way to Idaho Springs as part of Denver Mountain Parks' Circle Route G, later known as the Lariat Loop. The Peak to Peak Highway was soon planned to run from Long's Peak to Pike's Peak, including a section through Central City down Virginia Canyon to Idaho Springs then south over the ridgeline of Mount Evans and on to Colorado Springs.

Georgetown's boosters became frantic as these road projects seemed to exclude them entirely. Frustrated merchants in Georgetown wanted to make sure travelers found their way west, so they quickly erected signs along the road with a variety of somewhat wordy encouragements: "Good auto roads to Georgetown–Silver Plume. The Famous Loop and Lakes"; "The Great Gold and Silver mines of Colorado. At Georgetown and Silver Plume"; "Summer and Winter, Georgetown and Silver Plume excels in scenery and climate"; "Good trout fishing near Georgetown and Silver Plume"; "Switzerland's rival in beautiful scenery. The Auto road to Georgetown and Silver Plume"; "Green Lake the Gem of the Rockies, near Georgetown"; "Visit Georgetown and Silver Plume and their surrounding places of scenic beauty"; "Green Lake Scenic Road, one of the highest auto roads in the World, at Georgetown"; "The Heart of the Rockies—Georgetown and Silver Plume; One hour's drive to many hours of pleasure from Georgetown and Silver Plume"; "Good Roads to the Climax of Mountain Scenery. Georgetown and Silver Plume"; and "Good auto roads to Georgetown and Silver Plume, the Ideal Mountain

Cities."[23] When one of the signs disappeared, the Georgetown paper pointed the finger at the perceived "jealous" merchants in Idaho Springs, only to retract the statement the following week when it was learned that a property owner hadn't given permission for the sign to be erected in the first place.

Competition between towns reached a fever pitch when Idaho Springs' residents circulated a petition calling for an election to move the county seat to their town. The effort failed, but tempers flared, leaving the two ends of the county at odds with each other once again. Idaho Springs requested a redistricting of the county commissioners' districts since the bulk of the county's population had started to move toward the gold districts of Idaho Springs after the economic upheaval following the repeal of the Sherman Silver Purchase Act. The commissioners from Georgetown and Silver Plume, however, outvoted the single commissioner from Idaho Springs, and nothing was changed.

Even with new signs, the Booster Club realized that "autoists" required a destination—things to see and do. In order to reach Green and Clear lakes, drivers needed to maneuver along the sometimes treacherous, unnervingly steep road up Leavenworth canyon to Silver Dale. In a 1915 article entitled "LETS HAVE GOOD ROADS" the editors noted: "The ordinary auto cannot make the first hill to Green and Clear Lakes and to Argentine and Naylor Lake. We want this road put in shape so that these places can be reached by auto. A road could be easily made around the face of Leavenworth mountain which would cut this hill out and make the trip to the lakes possible."[24]

The preliminary review projected a cost of $3,000. One hundred Georgetown residents packed the halls of the county courthouse to request the new road. The Georgetown and Silver Plume commissioners voted to authorize $2,000 for the portion of the road south of the town limits, with the stretch across the face of Leavenworth Mountain to be built by the town. Georgetown's "boosters" immediately set to work and raised $867.50 in cash and pledges within two weeks. Mrs. Cora Seifried helped when she opened the Alpine Lodge in the elegant old Hamill residence on Argentine Street for a fundraiser, charging ten cents a head for a party with all proceeds going to the road project.[25] A ladies workday was planned for the end of the project, when all of the women in town would be asked to rake rocks to help smooth the surface for travel.[26] The ladies cooked and served meals for the volunteers as the road snaked its way up the mountainside.

The next project involved the neighboring town of Silver Plume, as plans for establishing an auto route across the old Loveland Pass came to the front. In 1916 the movement started to gain momentum: "In this day and age when the public is desirous of good automobile roads for the convenience of transacting business, or for seeking health or pleasure in these beautiful wonders of the Rocky Mountains, would it not be well for the citizens of Clear Creek, Summit, Park and Lake counties to seek sufficient funds to reconstruct the old Leadville stage road over Loveland pass for automobile travel?"[27] Once again the local workforce started the project, but the length of the road would overwhelm local resources for some time to come. The state legislature and state engineers would have to became players in the county road system before these major projects could be completed.

Community support for new and improved roads to and within the mountains reached an impasse in 1916, when a plan was brought forth to designate a national park centered around Mt. Evans. The editor of the *Courier* sounded the alarm when informed of the proposed Denver National Park:

> Are you aware that the proposed Denver park will include 15,329 acres south of Georgetown, Silver Plume and Idaho Springs in Clear Creek county? There are several beautiful lakes in the proposed park and many mountain streams in which there is good fishing, as well as one of the finest spruce forests in the West and much unexploited mineral land. All this will be taken away from the miner and prospector, and our mills whose tailings empty into Clear Creek may be stopped by injunction because some Denverite wants to fish. Not many years ago he would not have been able to buy a fishing rod and line had it not been for these same miners and prospectors.[28]

The fact that "some Denverite wants to fish" would become a double-edged sword for the residents of Georgetown. Tourism provided stable income for merchants longing for the business of previous years; however, the county was not willing to give up on its mining past.

During the first two decades of the twentieth century, the town's business community underwent numerous changes. Many long-term merchants left, while others assumed new products to fill the gaps. Samuel Strousse's dry goods store closed its doors in 1912: "Mr. Strousse senior thinks it about time to retire from active business, while Mr. Strousse junior sees a more profitable field on the western coast."[29] The same year, Kneisel and Anderson grocery store purchased the adjacent building to the west, adding to their interests the inventory of the Georgetown hardware company in 1913.

The Alpine Lodge (formerly the William A. Hamill family home) planned to bring European elegance to Georgetown based upon the Seifried family's visits to Europe. Plans came to a halt, however, with the death of Mrs. Seifried in December of 1916: "The funeral of Mrs. Henry Seifried occurred last Sunday at noon at the Seifried residence, with Christian Science services conducted by Mr. R. C. Vidler. The Order of the Easter Star of which Mrs. Seifried was an honored member, gave their impressive service at the cemetery."[30]

The First Church of Christ Scientist, Georgetown, came into existence in 1902, renting the building formerly occupied by the fledgling Georgetown Library on Taos Street. The small, yet enthusiastic, congregation soon purchased their own building at the corner of Seventh and Taos—a former saloon. Lecturers included the well-known Bicknell Young, "Dean of Christian Science Teachers."

During the early years of the twentieth century Georgetown supported six churches: Catholic, Christian Science, Episcopal, Methodist, Presbyterian, and Swedish Lutheran. The Catholic, Episcopal, Methodist, and Presbyterian churches housed ministers or priests in manses owned by the congregations, often donated by members who had left town. Youth groups included the Young Girls Friendly Society, the S. C. Club, the Little Helpers of the Episcopal Church, the "Cheerful Workers," and the Epworth League of the Methodist Church.

A new musical group, the Columbine Band, made its first appearance in 1908 sporting black uniforms with gold braid and an embroidered columbine, all made by Mrs. J. S. Fisher. After several stops and starts, and a call for music—"as a number of the important parts of some of the music is missing"—the band started playing at events around the area. By 1911, the group's repertoire included a number of well-known band arrangements:

1. March, "Faithful"
2. "Intermezzo Russe"
3. Waltz, "Lover's Devotion"
4. Alto Solo, "Come Where My Love Lies Dreaming"

5. "Silver Threads among the Gold"
6. Baritone Solo, "Senora Intermission"
7. "Indian Maiden"
8. "Old Church Organ"
9. Overture, "Sweet Briar"
10. Bass Solo, "Orpheus"
11. "When the Moon Plays Beekaboo [*sic*]"
12. "King Cotton"[31]

Many fraternal and social organizations remained active; however, those with multiple subgroups began to merge. The three area chapters of the Oddfellows—Georgetown Lodge No. 5, Harmony Lodge No. 13 of Georgetown, and Silver Plume Lodge No. 25—consolidated in 1915, keeping the name of the oldest chapter.[32] Georgetown Lodge No. 12 A.F. and A.M. absorbed the membership of the second local chapter, Washington Lodge No. 48, in 1916.[33]

Women's clubs included the Student's Literary Club and an Opera Club, while the former Shakespeare Club ceased to meet after the organization ended up in district court in an effort to solve a dispute over the election of officers.[34] In 1915, a new type of organization came to town when the Colorado Mountain Club planned a two-day hike from Georgetown to Grant. These clubs, along with the start of Boy and Girl Scouts, reflected the burgeoning interest in camping, hiking, and outdoor recreation that would grow throughout the twentieth century. These new travelers usually arrived in cars, camped under the stars, and explored communities and forests that continued to change as the forces of the twentieth century began to alter the landscape.

Edward Wilcox's dream, the Argentine Central, struggled to survive, as the cost of running a railroad that could only operate for a few months a year—and never on Sunday due to Wilcox's rigid religious beliefs—became overwhelming. Schedules changed (growing shorter), ownership and names changed, and doubts remained as to the future of the operation. In an ironic twist, William Rogers, Wilcox's adversary in all things related to Waldorf, purchased the line in 1912.[35] By 1915 most of the traffic along the tracks came with "auto-trucks" capable of driving along the rails. The following year a dance pavilion with guest cottages was built along the railroad grade on the face of Leavenworth Mountain, with guests walking or driving from Georgetown and Silver Plume.

The era of great mines, and the cantankerous railroads built to serve them, wound down in the early years of World War I. By the end of 1916, the townspeople were simply hoping for some kind of change and, unfortunately, were soon to get what they hoped for.

## PHOTO ESSAY 3

# THE YEARS LEADING UP TO THE FIRST WORLD WAR

*Construction of the hydro-plant at the east end of Sixth Street in 1900 provided badly needed employment and the promise of affordable energy for mines and homes. In 1906 United Light and Power merged with the Green and Clear Lakes Company and Cascade Electric to create United Hydro Electric Company. Denver Public Library, Western History Collection, Rocky Mountain Photo Company, x1029.*

*The corner of 6th and Rose streets c.1910 shows a plethora of telephone and electric lines that still couldn't keep businesses in town. Louis Cohen's COD building on the left-hand corner sits empty. Image courtesy of History Colorado (Mountain States Telephone collection, f40543).*

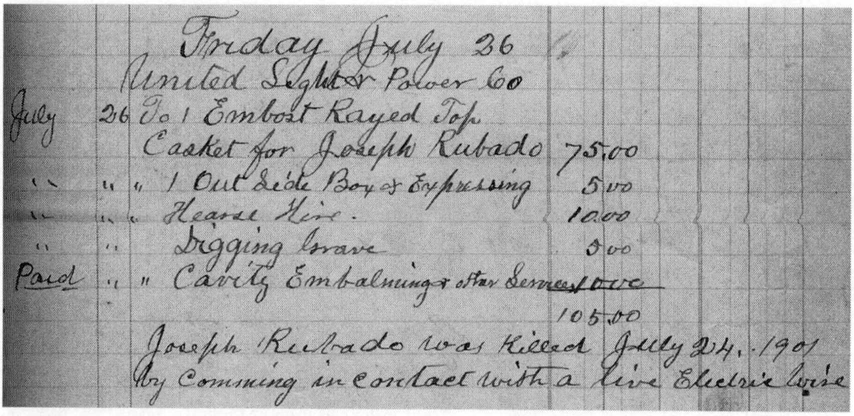

Friday July 26
United Light & Power Co
July 26 To 1 Embost Rayed Top
Casket for Joseph Rubado 75.00
" " 1 Out Side Box & Expressing 5.00
" " Hearse Hire 10.00
" " Digging Grave 5.00
Paid " " Cavity Embalming & other Services 10.00
105.00
Joseph Rubado was Killed July 24. 1901
by comming in contact with a live Electric wire

*The new, high-power electric lines brought danger with them. Joseph Rubado was electrocuted in July 1901. Henry Boyer's mortuary record book, Clear Creek County Archives, Georgetown.*

*The American Sisters Power Plant is shown on this photo postcard dated 1907. The plant was built by John J. White in 1906. In later years it would be known as the Pirie Plant. Bradley collection.*

*George Dalgleish took this photo of five serious, well-dressed gents in May of 1906. The pins on their lapels have three rings, which may mean they were members of one of the local chapters of the Independent Order of Odd Fellows, also known as the "Three Links Fraternity." The links represent friendship, love, and truth. Bradley collection.*

*The "Opera House" on Rose Street, also known as the auditorium, was purchased by Sarah Burkholder in 1907 and used for a variety of public events. Hotel De Paris collection, Colonial Dames of America, Colorado chapter, Burkholder collection.*

*This "surrey with a fringe on top" is headed up the old road to Green Lake. The woman on horseback is riding sidesaddle. Hotel De Paris collection, Colonial Dames of America, Colorado chapter, Burkholder collection.*

*This photo, looking north from Leavenworth Mountain about 1910, shows the tailings of the Centennial Gold Mine in the lower right-hand corner. Image courtesy of History Colorado f15288.*

*This photo of the Georgetown School, taken around the turn of the century, shows the young cottonwoods still protected by boards to keep livestock from damaging their bark and provide support in case of strong winds. Denver Public Library, Western History Collection x1127.*

*Men and horses gather at the corner of Sixth and Rose Streets after the record-setting snowstorm of 1913. Denver Public Library, Western History Collection f47255.*

*The Star Hook and Ladder team of Georgetown pose for a picture after winning the race in Idaho Springs in 1910 with a time of 24 14/15 seconds. Courtesy Georgetown Volunteer Fire Department.*

*The members of Georgetown's Columbine band pose with "Pop" Olds of the Fire Department. Denver Public Library, Western History Collection x6532.*

*The county jail stands as an imposing building on Biddle Street in 1912. Denver Public Library, Western History Collection x1292.*

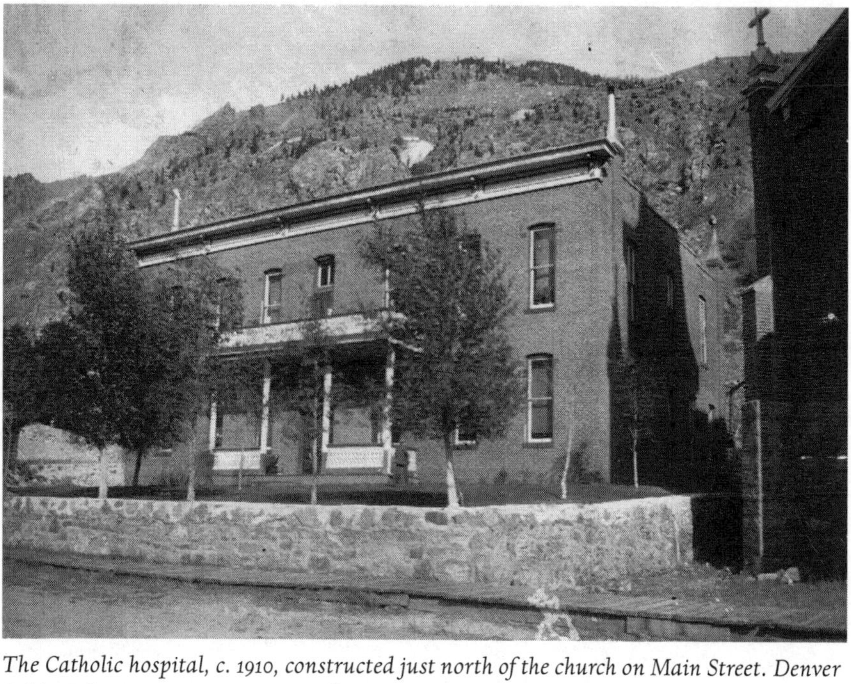

*The Catholic hospital, c. 1910, constructed just north of the church on Main Street. Denver Public Library, Western History collection x1094.*

*This photo of Georgetown was taken from Democrat Mountain c. 1910. The Capital Prize Mill and Catholic church on Main Street can be seen in the distance. Photo courtesy Archives, University of Colorado at Boulder Libraries, McLean photo 1277.*

*Well-defined streets, sidewalks, and trees in a city park—all the signs of a permanent community can be seen in this view from Tenth and Rose Streets looking east about 1910. Photo courtesy Archives, University of Colorado at Boulder Libraries, McLean photo 1276.*

*The hillsides are barren, but trees are starting to take hold on the town streets. Photo taken about 1910 near the corner of Ninth and Rose Streets looking east. Photo courtesy Archives, University of Colorado at Boulder Libraries, McLean photo 1279.*

*The sign on the back of a building near the Centennial Mine would have been visible to all who traveled by train to and from Silver Plume. Denver Public Library, Western History Collection x1354*

# CHAPTER 4

# MINING WHAT THE MOUNTAIN HAS TO OFFER

As the profits from precious metal mining faltered, dreams of other prospects and new minerals lingered, including new mining machinery. With electricity available throughout the district, new, or improved, machines soon appeared. These included a variety of machine drills, including the highly promoted Sigafoos tunneling machine. Georgetown even got in on the act when Starky Jones invented a "lubricating valve," which "proved itself to be a good thing."[1] Each one promised improvements and/or profitability; some actually worked. Others proved only straws of desperation.

One prospect remained, finding a new and profitable mineral to mine. For a fleeting moment, one had appeared, molybdenum, a "very hard metallic element," eventually found to toughen steel alloys, among other uses. Back in 1890, "good specimens" were mined "occasionally" in the American Sisters and other mines south of Lawson. Little came of them, at the moment, as the paper admitted. Miners seemed "puzzled over what it is, and finding out what it is, are still in the dark." All the *Courier* could tell its readers was that "molybdenum is a white metal, brittle, and difficult to melt." To the editor, it did not seem to be of much value. "Its only use is in the manufacture of compounds used in the laboratory."[2]

"Moly" might not have had a profitable use then, but the time was coming when it would. So that hope momentarily vanished in a twinkling. Meanwhile, Clear Creek County continued to produce lead, copper, and zinc, none as glamorous or exciting to the general public as gold and silver. Only once prior to World War I did copper production top $100,000, and that was in 1916, before the American entry into World War I. Lead, however, as a byproduct of silver mining, hit a peak of over $300,000 in 1899. Only twice did it slip under $100,000 in the same period.

Zinc, as uninteresting as it was to the public, kept Clear Creek County in the headlines. Josiah Spurr had predicted just such a development in his "Outline of Results" in the *Economic Geology of the Georgetown Quadrangle*. There appeared, he felt, "to be a good outlook for zinc production in the future."[3] After 1903, the mines in upper Clear Creek became large producers of that metal. Smelting problems limited production until a method to refine it came with, among other innovations, the Wilfley table used to separate minerals. By the early 1900s, mines were producing over a million pounds of zinc, a figure that held steady until the World War I era. The value, though, reached only a high of $344, 000 in 1916. Value excited people, not pounds.[4]

In the years of the early twentieth century, Georgetown silver production continued steadily downhill. Indeed, by the outbreak of "the Great War," as they referred to it, silver mining was about half of what it had been at the century's turn.

From reading mining reports and local newspapers, that reality would have been somewhat hard to grasp. For instance, the Independent Mining Company solicited investors with a promotional booklet describing the glittering potential of the Griffith lode, "one of the best possibilities in our camp." That booklet even looked like one of its forebears from a generation ago, with promises, promotion, and a dash of mining terminology.

The *Georgetown Courier* had been encouraging just such a project for years. In its December 28, 1901, issue, it praised the old Griffith Mine, which was once again being developed after it had been "abandoned over 20 years ago." Then, the editor went on to complain: "All of these veins were known to exist but the remarkable apathy that has characterized the mining operations here for some years past has been the chief cause of their lying unopened for so long. With the advantage of new blood and energy, backed by capital to do the necessary development work, great advancement has already been made and the future holds promise of far greater." This year-end summary of mining concluded that "as a district offering opportunities for investment in mining there is no better in Colorado."

Note the lure. The mine never failed. The "old-timers" simply suffered from "remarkable apathy." (One wonders what they thought about this interpretation of the past!) This lethargy contrasted clearly with the "wide awake," energetic newcomers who would ride to the rescue.

The Independent Mining Company did not miss a beat. Among the mandatory photographs, maps, assay reports, and cross-cut drawings of the mine—came a history of the Griffith Mine that paid "from the grass roots," the company reassured its readers. "The mining industry is now upon a basis similar to that of every well established industry. One in which the facilities employed determines the extent of operations." To further drive that point home, the prospectus went on to boldly state: "as such it can scarcely be called a prospectus, for an endeavor will be made to avoid the discussion or prospective conditions and limit this statement to a recital of *what we have and how we will get it to market*."[5]

That being said, the pamphlet continued on: "To the experienced mining man it is apparent that the mining districts of Clear Creek County were made to be worked by tunnel." That was precisely how the company planned to do it, but wait—the author was not finished—readers were further assured, "The ores are transported by gravity to the level of the railroad more cheaply than to hoist from shafts and haul to market in wagons." And, finally, "the operations of the Griffith mine and other mines of the group by laterals upon the several veins from the main tunnel illustrate every principle that distinguishes scientific mining of today."

Add to all these benefits the railroad that provided year-round transportation "from the portal of the mine" only fifty miles to Denver with its "up-to-date" smelting processes. Further, water-generated electric power, which cut costs and eased mining, was available at the mine, and "last but not least," combining all of these advantages with tunnel mining, they produced "the elements that have changed mining from speculation to a substantial industry."[6] Every investor wished to hear such glittering promises. The company probably thought that indeed a sucker was "born every minute."

It seemed impossible that the Griffith and the other mines included in the company's holdings would not bring Georgetown back into its earlier bonanza days. The old adage still held true, "let the buyer beware."

News also appeared from all the other districts of the upper Clear Creek region. Silver Dale had its Colorado Central Mine, which made occasional appearances in the *Silver Standard* and the *Courier*. The pattern looked familiar. After being operated under lease, the company was reorganized in 1907, started tunneling, and exposed "a good streak of ore," with shipment making "profitable returns."[7] Silence, then, until 1916 when "this great property which has been idle for seven or eight years is about to be reopened."[8] The dumps were being sampled and, finally, in December, readers learned that the directors and stockholders "were more than satisfied with the results of the exhaustive test which have been made of the dump material."[9]

They now planned to build "a big mill." Meanwhile, the company intended to work from the tunnel to open both new and old areas. The company held a five-year lease on a property that had been worked for almost fifty years. The only time it was shut down was between 1907 and 1916. "The evident reason of the shutdown was the panic which swept the country in 1907, the effect of which lasted for several years following."[10]

The panic of 1907 had not been that long lasting nor damaging in general to mining. Why it had so affected the Colorado Central Mine, the article did not say.[11] It certainly, however, proved a convenient scapegoat.

Silver Plume paralleled its neighbor Georgetown in its early twentieth-century excitement over mining. The *Silver Standard*, for instance, in its May 30, 1903, paper, boldly proclaimed, "This seems to be an era of mill building for Clear Creek county as well as tunnel driving." Always upbeat, the paper continued: "These mills are of course erected for the purpose of treating ores that will not pay to ship to the smelters as they come from the mine, and it indicates the attention that is being given to the handling of the low grades, and while many of the high grade properties will continue to turn out the usual amount of smelting ore, a large increase in the production of the country will result from the handling of stuff that has heretofore been unprofitable." The author of this long sentence only echoed what his neighbors and readers believed was happening around them.

In the years that followed, the paper reported on the usual exciting mining news items—electricity installed, operations resumed, mill buildings commenced, tunnels dug, new machinery introduced, and leasers who opened up new discoveries. With less enthusiasm, the paper reported on operations suspended, equipment not working right, and a fistful of other negative matters.

Not that the Silver Plume folk gave up. It was far from that. Their newspaper folded in 1907, but the *Courier* carried forth their banner. The September 2, 1916, paper headlined "Boom at Silver Plume." The editor opened with his appraisal of the situation: "to one who has not visited Silver Plume for a few months the aspect of the town and the mines surrounding it is such that it takes some little time to comprehend the immense nature of what is today an exceedingly busy change." The article continued, "several hundred men are at work in the mines and what insures the great growing prosperity of the place is that nearly all the men are actually engaged in producing pay ore."

Between Silver Plume and Georgetown, near the tracks of the Colorado and Southern, sat the Lebanon group of mines, operated years before by a tunnel. The Lebanon serves as a classic study of the district's hopes post-1900. The *Silver Standard* (December 5, 1903) announced that local rumors hinted the mine would be put into shape for production. This stirred the editor to point out a lesson to be learned. "The property has lain dormant for some time on account of disagreement among the owners, but it has a good record, and no doubt only needs to be put in shape for working to repeat the history of many other properties in this district which have been taken hold and by the judicious expenditures of a little money have become profitable." Two weeks later, he noted hopefully that the work of cleaning out the tunnel "is now going on." A "bad cave in" occurred a number of years ago, which had to be cleared out and the tunnel retimbered.

Ever encouraged, the same paper reported, in September 1904, that work was "getting things in shape for bringing this once famous property back into the list of producers again." Ten lodes had supposedly been cut into while driving the tunnel ahead, and "others lie in the line" of the tunnel. Once again, the editor reminded his readers, this was "one of the many good propositions in this district which only need the expenditure of a little capital in development work to yield up their hidden treasures."

The next April, readers found out that the mine was operating under an option and that half the property had been sold. Again, they learned that this "will aid in restoring the old workings for future development." A planned mill would allow the profitable mining of low-grade ore, "as soon as [the mine] is put in shape for carrying on extensive operations."

Over a year later, the lessee of the property had brought in power drills, and indicated that soon a larger force of men "will be put on." Trial shipments of ore satisfied the lessee of the mine's value, and he now worked on stockpiling ore to have a good tonnage "at hand." Optimism still permeated the air. Ahead "of the main tunnel there lies a very rich mineral country which so far has merely been scratched on the surface," or so the lessee hoped. Meanwhile, the Lebanon was rumored to be consolidated with a neighboring property that would "result in a splendid group of claims which in early days were large producers."

Alas, despite such optimism over the past three years, the Lebanon never attained its "assumed" wealth and prosperity, having been "idle for over twenty years" before this revival.[12] The Lebanon's rebirth would have to await another day and time.

Another relic of the past also made a return appearance: the Stevens Mine, once partly owned by George Armstrong Custer, in his grave over a quarter of a century. The *Courier* introduced the new operations of the Stevens by presenting a familiar refrain: a historic summary of its "storied past" and how "modern methods" would make it profitable. In October 1903, the *Silver Standard* announced that a concentrating mill would be built to handle low-grade ore that could not be profitably worked or shipped "during the many years that it was so successfully operated." That statement proved a bit of a historic stretch and a dose of wishful thinking. The Stevens had not been very profitable for Custer and his contemporaries.

The reopening of this property did not sit well with Silver Plume. "There is some apprehension on the part of the citizens of this place that the operation of a mill at the Stevens mine will interfere with the water supply of the town." The article declared that it would make the water "so roily as to render it unfit for domestic purposes."[13]

Concern over mining's impact on a community's water supply was not new. Other Colorado towns had also raised the issue. Fundamentally, however, it posed a difficult question—a prospering economy versus a threat to public health. For a while yet, mining would prevail, barring a major health crisis.

The Stevens reopened, and, by July 1904, was "running steadily night and day and treating about 50 tons of material each 24 hours from the old stopes of the mine." More machinery would soon be installed to increase the output.

The water issue would not go away, however. In July 1905, Silver Plume's town council met in "special session" to consider the "water question." The rather arrogant manager of the Waldorf Mining Company, which operated the Stevens, E. J. Wilcox, bluntly told the council that he declined "to go to any expense to prevent a pollution of the water supply, such as now renders the fluid unfit for domestic use." Not stopping there he "intimated that if the town desired to test the matter in courts he was ready to have it commence an action." Such a bullying, arrogant attitude did not speak well for Wilcox or mining's image.

He then fell back on an argument that many of his contemporaries used. According to the reporter, "he stated that he had complied with the requirements of law and seemed to feel aggrieved that the board should call his attention to the fact that the water is being polluted, intimating that they were antagonistic to the investment of capital in mining enterprises."[14]

Council members visiting the Stevens pointed out that the settling pond, constructed to alleviate the problem by "filtering" the water, "was not only full of tailings from the mill but that not all the tailings were being run into it in the first place so that it was practically of no value for the purpose it was intended for." All this almost sounded like an argument presented by a later generation. Neither side won, despite threats and counterthreats. Eventually the Stevens shut down, thus eliminating some of the pollution. Then the Stevens faded from the news, to be replaced by other promising properties.

Meantime, both Georgetown and Silver Plume papers continued to boost their mines and mining. The *Courier* was active, carrying a mining column on the front page under such titles as "The Week's Story of the Mines." It praised the interest shown by eastern investors, new ore treatments, and an individual mine's progress, and it gave a year-end summary. It also applauded the economical water power and tunnel system, the short distance to the Denver smelters, and ease of railroad transportation. History also came to the forefront to bear out the wonders of the old days, which could be revived.

Always optimistic, the *Courier* of December 27, 1902, for example, noted, "the confidence of the operator in the long neglected mines of this district is the result of past experience which demonstrates that our veins descend to the greatest depth." Soaring into poetic rhetoric, the paper continued:

> The abandoned shafts of these pioneers dot our ranges, like the clusters of the Milky Way. The visitor, inexperienced in mining regions, may be prone to regard these numerous shallow shafts and short tunnels as evidence of modern failure; . . . such inexperienced people forget that of such holes could be made mines now within a few months, at a comparatively small cost.

Much to locals' pleasure, the district received an added boost when the prestigious *Engineering and Mining Journal*, in its year-end summary of Colorado mining on January 3, 1903, stated that "on the whole, it can be said the best outlook is in upper Clear Creek, and in the San Juan region." It was here that "young mines help redress the balance of old exhausted properties." And it came right back the next year (January 7, 1904) to praise the progress of upper Clear Creek.

"Always next spring!" Mining matters would involve a statement that any baseball fan could completely understand. Another longing never died, either here or throughout Colorado. Denver's *Weekly Republican* of January 1, 1910, expressed it beautifully. Many camps "have been in a measure marking time and waiting for silver to come again into something near its own." A higher price would bring back the "bonanza."

Another sign of Georgetown's changing times came with the publication of the annual *Colorado State Business Directory*. The 1913 volume, for example, listed only three mining engineers and one man identified vaguely as "mining," along with a handful of mining companies headquartered in the community. That proved a far cry from what would have been found a generation earlier.

One thing never died—that was optimism. Again, the *Weekly Republican* of January 1, 1908, caught the spirit. Miners "could always stand to have things a little better, but time has taught him to take them as they come. He has his eyes glued on the future."

Setbacks did not bother this individual. "He knows that he has the greatest mining field in the entire world [Colorado], the biggest and richest storehouse of mineral to draw from. The best conditions for mining and living in the meantime."

The "outlook" might "never be brighter," but the second decade of the new century brought forth no bonanzas. Leasing continued, some old prospects were reopened with much fanfare, and work "resumed in old rich lead and silver" mines. Georgetown and Silver Plume, however, could no longer claim to be newsmakers around the state and nation.

Idaho Springs emerged solidly with its gold mines as the county's "hot" district, but even there, the old days did not return. Clear Creek County had seen better days, as 1916 dawned to a world far different from a generation ago—or even a decade.

Both the old and the new came with the New Year. As 1915 ended, the *Courier* poignantly remarked that if silver would be "fixed say at 75 cents per ounce, the mind of the silver miner would be at rest." It could be 1893 all over again, as far as the editor was concerned. Another issue agitated some miners' minds. Prohibition was now the law of the state, as the paper pointed out on January 1: "Georgetown is the driest town of its size in the state. Even a doctor's prescription is of no avail." That old bulwark of the mining west, the saloon, disappeared also. Yet readers were reminded of the dangers of mining during the winter in the very next issue, when three men surveying mining claims were swept away in a snowslide and two were killed.

The new world clobbered Georgetown's mining in ways that hurt, both financially and emotionally. The American Smelting and Refining Company, a monopoly if ever there was one, had gained control over all the major smelters in the state and elsewhere soon after the century's turn. In the years that followed, the company went about setting smelting charges, and, seemingly, little existed that the mine owner could do about it. What had happened to the freewheeling days of the independent miner who was master of his own destiny?

Protests appeared quickly. Nevertheless, from the miner's perspective the situation only got worse. Complaints mounted, and Clear Creek County, by 1916, reflected the mood of the times. In March and April, a flurry of meetings of the county branch of the Colorado Metal Mining Association aroused the populace and encouraged it to join the organization at a most reasonable fifty cents per year in dues. Meetings in Georgetown and Silver Plume expressed miners' concerns, fears, and wishes.

The organization wanted the government to do something to reduce the "trust's" power and price structure and "improve mining conditions generally," while it challenged local miners to "work and stick together." When the trust made a $13 million profit in 1915, locals complained, "why don't we pause to consider that 45% of our lead, 50 percent of our copper and from 60 to 80 percent of our zinc went to swell that 'net profit.'" This "cause is of the utmost importance to its [Georgetown's] very existence," the editor ardently argued. "Is there any reason outside of 'because we can,'" that the company could take such a highhanded approach?[15]

The fury came to naught and by the end of the year the slogan of the hour indeed proved true:

> Miners may come and miners may go,
> But the Smelter trust goes on forever.[16]

To some it seemed a war for Georgetown's and Silver Plume's survival, to others the world news seemed much more threatening. As the "Great War" dragged on in the trenches and the slaughter mounted, 1916 came to an end; with it Georgetown's era as a front rank Colorado mining district finally drew to a close – a reality that even locals in their truthful moments had to admit. Obviously, it had happened years before, but few willingly admitted to such a gloomy assessment.

Georgetown had evolved over the previous decades from the excitement of new discoveries, through the despair of trying to find a reduction process that worked economically, to the high tide of its boom, with the appearance of large companies. Then it went into a gradual decline, ribbed with hopes and disappointments, as small mine owners struggled to survive and continue mining.

Clear Creek County definitely had reached the last stage. Signs existed everywhere, from the high mountain valleys to the streets of Georgetown and Silver Plume, to the wind whistling through abandoned buildings in the small mining camps that themselves stood with one foot in the grave. They might cry out against their fate, but few on the outside heard or, if they did, paid much attention.

In this pattern, Georgetown offered no exception, following a familiar historic course in the mining West. What the future might hold, the community did not know, but there lived in the town and district people stubbornly clinging to the belief that the best mining days remained yet to come.

The exciting times of 1859 and 1864 for Colorado and Georgetown had receded into old-timers' fading memories and tourists' imaginations, as they prowled around abandoned mines and desolate camps where once life had throbbed. Still, mining dreams flickered ever so faintly among the faithful.

It all seemed like one of those newfangled motion pictures with their "Perils of Pauline" scripts: the heroine on the brink of disaster, when finally she was rescued and goodness triumphed. What Georgetownites wanted more than anything was a happy ending to their story. Then, they could all live happily ever after, just like in the movies.

PHOTO ESSAY 4

# MINING IN AND AROUND GEORGETOWN AND SILVER PLUME

*Philip Poirson's eight-horse team stands at the corner of Rose and Sixth Streets in Georgetown, taking a new compressor to the Wilcox Tunnel near Waldorf about 1905. The team would travel twelve miles south, gaining 2,000 feet in altitude. George Dalgleish photo, Bradley collection.*

*Labeled "Main Street, Silver Plume," this George Dalgleish photo shows mining equipment headed east, probably to the Ashby or Payrock Mines. The reason for the children on donkeys and horses is unknown. The mustachioed gentleman in the center of the street, with his young son aboard a donkey, is Italian immigrant Peter Sibelia. Denver Public Library, Western History Collection x2180.*

*The Dolly Varden Mine sat perched on the side of Saxon Mountain above Georgetown. The mine operated around the turn of the century, with great views but a long and difficult path to haul ore to a mill. Bradley collection.*

*This photo of the Payrock Mill (on the far eastern end of Silver Plume) shows a wide variety of mining equipment, along with some solid rockwork in front of the mill and across the creek where the tall rock wall secured the railroad bed of the Colorado and Southern on its way into town. Denver Public Library, Western History Collection, x61593.*

*The Diamond Tunnel property above Silver Plume hosted a few large picnics as well as tours for visitors and investors. The number of women in this party would have worried those who believed that women were bad luck underground. Colorado College Special Collections, Kernochan, pp. 82–22a, album 3, page 94.*

*Hand drills and miners' candles are shown in this photo taken in the Cosmos tunnel of the Doric Mine in Georgetown. The Doric produced a limited amount of gold around the turn of the century. Interest in these claims resumed after the price of silver ore dropped after 1893. The photo shows young Bert Snodgrass in the middle of the picture, along with W. J. Lewis Sr. and Jr., and others in the background. Denver Public Library, Western History Collection, x61609.*

*The second mill of the Colorado Central mining company, built on the foundation of the earlier, much larger facility. The buildings at the mouth of the Marshall Tunnel can be seen in the back right, tucked in at the foot of the mountain. The road from Georgetown ran along the base of the mountain in front of the smaller buildings. The photo would have been taken about 1900. Denver Public Library, Western History Collection, z6848.*

*This view of the Burleigh Mill on the west end of Silver Plume shows several adjacent buildings and the large smokestack that was put up in 1905. Chappel family collection.*

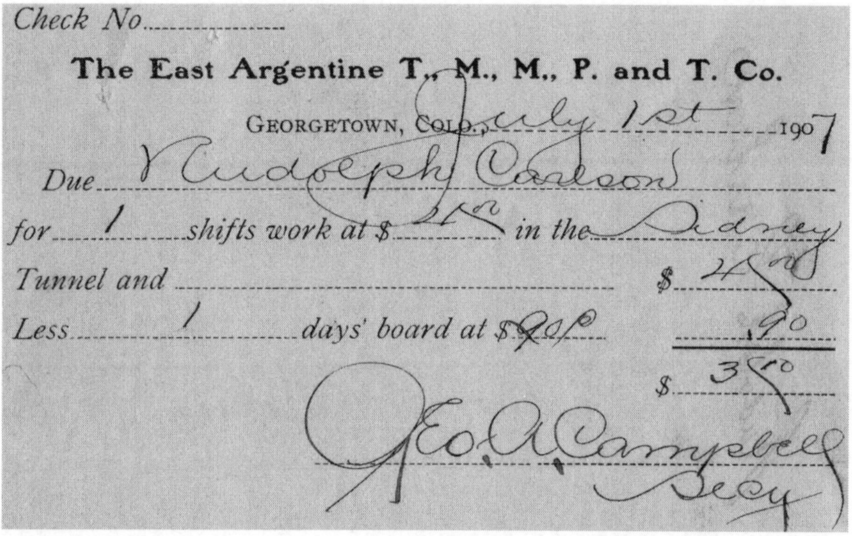

Check No..................

**The East Argentine T., M., M., P. and T. Co.**

GEORGETOWN, COLO., July 1st 1907

Due Rudolph Carlson

for 1 shifts work at $ 4.00 in the Sidney

Tunnel and $ 4.00

Less 1 days' board at $.90 .90

$ 3.10

Geo. A. Campbell
Secy

*This paystub from the East Argentine Transportation, Mining, Milling Power and Tunnel Company, Waldorf, in 1907 shows wages were $4.00 per day less $.90 for one day's board. Bradley collection.*

*This Dalgleish photo shows the Burleigh Mill, looking west, with the smokestack pouring smoke. The building burned to the ground in the 1920s. This may have been the experimental mill used by the Edison investors around the turn of the century. Image courtesy of History Colorado (George Dalgleish, photographer, 10031975).*

*Nashotah (also spelled Nashatah) mill at the foot of the 7:30 road above Silver Plume was completed in 1907. The large tailings pile up the mountain belongs to the Diamond Tunnel property. Denver Public Library, Western History Collection, f40303.*

FACING PAGE. *Miners at the Capital Prize Mine in Georgetown take a break with lunch pails in hand about 1915. 1. Murray Bruce, 2. Walter Oklun, 3. Albert Oklun, 4. Homer Hamilton, 5. W. E. Walthers, 6. James Barnes, 7. James McUlser, 8. Fred Oklun, 9. John W. Smith. Walter Oklun later served as a private in the United States Army in World War I, succumbing to influenza in France on October 5, 1918. Denver Public Library, Western History Collection x60928.*

*Wilfley tables inside the Nashotah Mill were used for concentration of ore. Several electric lights helped illuminate the work tables. Denver Public Library, Western History collection x60714.*

*New structures were built at the foot of Democrat Mountain with renewed interest in gold mining at the turn of the century. The Kelly tunnel headquarters and power house are shown in this photo c. 1910. Denver Public Library, Western History Collection x61646.*

FACING PAGE. *The "new" Griffith Mill, constructed in 1906 well after the brothers had left town, was built north of what is now called the north tunnel of the Capital Prize Mine along Biddle Street. The tunnel accessed the workings of the East Griffith, later combined with the holdings of the Capital Prize. The building was later torn down, moved, and reassembled as the Centennial Mill in the south end of town in 1929. Denver Public Library, Western History Collection, George Dalgleish photo, x61655.*

*This building housed the power house for the Kelly Tunnel after the turn of the century. The power house was along the creek off of Argentine near Seventeenth Street. The tunnel is west of Georgetown in the vicinity of the town shops. Denver Public Library, Western History Collection, George Dalgleish photo, f48361.*

*The third Colorado Central Mill in Silver Dale is shown in the background of this photo. Construction started in 1917; production started in 1918; bankruptcy was filed 1920. The building in the back right was the rock house, or crushing annex 73′ × 90′. The larger building in front was the concentrating plant, 113′ × 136′. James G. White photo 5192 taken in June of 1920. Charlie Wright collection, John Tomay Memorial Library.*

## CHAPTER 5

# LONG EVENING INTO NIGHT

The guns of August 1914 plunged the world into war, and try as he might, President Woodrow Wilson could not get Americans to be "impartial in thought as well as action," nor eventually "neutral in fact as well as in name during these days that are to try men's souls."[1] Loyalty to the old county proved too strong for many. These ties mainly supported the Allied side. British propaganda swayed others. Germany became the barbaric Hun, particularly with submarine warfare, but perhaps the strongest factors were the economic relations between the United States and the Allies, particularly England. By 1917, few Americans remained impartial.

Far off in Georgetown and Silver Plume, newspaper readers were amazed at what quickly transpired in Europe. Such new weapons as tanks, gas, machine guns, and submarines horrified them, though probably at the same time, they were being fascinated by airplanes, as they soared over enemy lines and fought each other high in the clouds above the trenches. Terror and death lay on the war battered ground below them.

War, though, has a ravenous appetite for base metals, such as copper, lead, and zinc, to produce those needed military goods. Mining then would surely benefit in such a tragic situation. Actually, the whole of the American economy benefited, as the war dragged on through 1915, 1916, and into 1917. Farmers, industrialists, and businessmen had goods to sell, and bankers had money to lend.

The reader may recall that production of Clear Creek County lead, copper, and zinc reached new highs in 1915 and increased again in 1916. Indeed, copper and zinc set all-time records. All of this reflected war related demands.

It all must be taken in perspective, nonetheless. Despite such records, Clear Creek remained a minor player. Lake County produced ten times the amount of zinc and five times the amount of lead, and San Miguel County three times as much zinc. Clear Creek, however, did outproduce neighboring, and longtime rivals, Gilpin and Boulder Counties. Regardless, it was all relative. Not one of them emerged as major producers of these base metals on the American scene.

Molybdenum, however, finally came into its own; the discovery of a use for moly, to toughen steel, made it a necessity to build up the country's military arsenal. Boulder County prospered, particularly in and around Nederland, but Clear Creek enjoyed its own excitements.

The *Georgetown Courier* (September 29, 1917) could hardly contain itself, headlining, "Extensive Work on Molybdenum Mines." It sounded like the old days. New roads were being built, tunnels dug into the mountains, timber cut for mines and buildings, aerial trams planned, mills under construction or being updated, and "many more men are required here at once."

The big news came from the Urad Mine above Empire and slightly off Berthoud Pass. County mining news such as this hadn't excited locals in years.

> The actual mining of this metal, which is most essential to the interests of Uncle Sam and his allies towards winning the world war now in progress is not yet under way, as the amount of preliminary work to be done is utilizing all the men, carpenters, teamster, sawyers, loggers, chopper, and laborers it is possible to get hold of.
>
> The grading for the big mill has been finished, but grading is going on for large settling tanks for the tailings so that the creek will not be contaminated.

> An aerial tramway will be built from the upper workings to the mill, so that the big dumps there can be concentrated. The milling consists of fine grinding and flotation.
>
> Many more men are required here at once.[2]

It all took place in the name of patriotism, of which Americans generated a superabundance during 1917–18. "The mining of molybdenum is one of the greatest and most important work that civilians can take part in to help Uncle Sam and his allies win the war. The preparatory work at Urad will continue well into next summer."[3] Molybdenum now became all the rage. It would bring the county and its communities back into the old days of prosperity.

It sounded like the old days in another way, when a night snowslide buried "20,000 feet of mill timbers" and about ten tons of coal. Fortunately, no one was killed, as the "site of the mill is located out of danger from snowslides." The lumber, unfortunately, had been stacked "where slides are frequent." Nonetheless, "that the slide buried the lumber out of sight, indicates that there must have been quite a bunch of snow, and it will be no little job to recover the lumber."[4]

By March 1918, the *Courier* reported the mill in operation, with a 200-ton daily capacity, production underway, and a large boarding house completed. That must have been a relief to the men sleeping in a tent just beyond that December slide.

More buildings were being constructed or planned. The *Courier* reporter said that those included "a large hotel which will be located near the mill to accommodate the company's employees." No one of "questionable loyalty" would be working on the property. "It is understood that the U.S. Government will take the entire production of the mines for use in the manufacture of munitions of war, so that the company is exercising great care in the selection of employees."[5] Americans throughout the country stood foursquare on the watch for unpatriotic people, saboteurs, and others who might undermine the war effort. So the mine was right in step.

Timing was essential at the moment. Finally and fortunately, the war neared its end as summer slipped into fall. Peace came at long last, on November 11, 1918, when the armistice brought an end to "the war to end all wars." As the world demobilized, in the following months, the molybdenum military market collapsed just as the Urad Mine came into full production.

Not that an industrial need was lacking. That would come in peacetime with the soon-to-be-thriving American automobile and steel industries. However, the Colorado molybdenum market fell on hard times from cheaper, foreign competition, the end of military and government needs, and a shrinking national market. The good time prospects vanished when the guns fell silent.

Reports from Urad in 1919 and 1920 reflected this. In July 1919, "on good authority," it was reported that a force of fifteen to twenty men would be put to work in development. "No attempt will be made at present to produce ore or operate the mill. Work was discontinued on this property about two months ago."[6]

The next January, this tidbit appeared in the *Courier*: "Rumor says that work will be resumed on the molybdenum mines at Urad within thirty days." Then the Vanadium Corporation of America purchased the property.[7] Developmental prospects vanished, but Urad was not finished as a molybdenum mine. Quite the contrary, it would eventually develop into one of America's larger producers. For the moment hard times settled in, and competition appeared. Colorado competition emerged from the Climax Molybdenum Company's new mine on nearby Fremont Pass that sat on top of one of the world's largest moly deposits.

Hard times settled in other ways as well. Lead, copper, and zinc had all helped lessen somewhat the collapse of the precious metals industry in the nineteen-teens in Clear Creek County. With the war's conclusion, however, base metals entered a volatile market, with the general price trend being down.

Only silver enjoyed a price revival, and that came about because of the war. The price of silver jumped on the world market to more than a dollar an ounce, particularly because England used silver to purchase war materials. "By the early part of 1918, Great Britain had exhausted its silver supply and called upon the United States for financial assistance."[8] Meanwhile, for years, Nevada senator Key Pittman, along with other western silver state senators, had spent a great deal of time fighting to pass a law to raise the price of silver to a point where mining it would again be profitable. Seeing his opportunity, Pittman sponsored an emergency measure (often called the Pittman Act of 1918), which provided for a loan of silver bullion to Britain.

The act, in effect, guaranteed the silver miner a minimum of a dollar an ounce for four years. The silver price reached a high of $1.12 in 1919. The expected result for Georgetown was the hoped-for arrival of a slice of boom days once more. Hopes died aborning. The value of silver mined went up ($400,332 in 1919), but, discouragingly, the ounces of silver coming from Clear Creek County continued drifting downward. Still, even that peak figure represented but a fading shadow of what the county had produced a generation before. Within two years, production fell to less than half of that figure, and the one hundred-thousand dollar range became the norm for a brief while, before it slipped further downhill again.

Precious metal mining limped through the war at this reduced rate and entered the 1920s without much real hope. Base metal zinc did the best of all Clear Creek County mining, ranging over a million dollars per year production, both before and during the war. Even it had slipped into the $200,000 range by 1924.

Without question, though, despite such stark reality, expectations continued to be as optimistic as ever. The Hall Mine on Leavenworth Mountain made headlines in late 1917 and held attention for about two years. It planned to cut lodes inside the mountains and "secured bonds and leases on nearly" all of them. Again, the old story appeared about earlier miners coming in, in this case in 1887, and making "absurd" decisions which led to a "season of Irish dividends" and the closure of the property. The new owners would do better, it was optimistically believed locally. They did not, and the mine closed.

The closure of a property might leave its miners with lost pay and only the courts as a recourse to recover their wages. By Colorado law, they had the first claim, but with no company assets, there often appeared to be little that miners could do. Unscrupulous owners or manager often missed paydays just before closing.

Men working at the mines received wages depending on their occupation and skill. Drill operators, timbermen, and blacksmiths earned $4 for eight hours, muckers who loaded the ore into cars and brought it to the surface and millmen made $3.50, while general miners secured $3 an hour. Some men actually took a gamble and were not paid at a daily rate, but rather were promised pay based on a percentage of the sale price of ore.

These wages were not noticeably higher than those of a generation before. Although they had dropped to under the standard $3 per day during the dark days of the 1890s, by the century's turn, they had returned to that average. Little else had changed. Despite improvements in safety, mining remained as dangerous, physically hard work. In the face of all this, men unrelentingly scratched around the surface and dug into the earth looking for hidden treasure.

The Colorado Central Mine, in Silver Dale, continued to raise hopes and create newsworthy items for the local paper, just as it had done before the war. As the *Courier* pointed out in its November 10, 1917, issue: "Colorado Central a Busy Place." Employing fifty men, the company was driving a tunnel, reopening and retimbering another one, and building a mill, and it had just finished a sawmill and construction of a tailings flume from the mill, "about a mile long," to the valley.

The war benefited the property. The chief ore values were "silver and lead, both in demand at present, and at prices which compensate for the advance in overhead charges since August 1914, when the world war was projected by the misguided Kaiser."

Reports continued to be good. In February 1918, an article in the *Georgetown Courier* was headlined: "Stupendous Work at Colorado Central." It went on to state that "anyone visiting the Colorado Central company's property will be impressed with the great amount of work which has so far been accomplished and with the extensive construction which is under way." In all, it hailed the company's effort: "It has taken considerable time, money and efficient brain work to gain the point at which this company stands today. . . . Therefore it is a very pleasant thing to know that the reward for these strenuous efforts and perseverance is at hand and during the coming summer we expect to see the enterprising men who compose this company, commence to reap [their] first reward."[9]

Even the *Engineering and Mining Journal* praised the Colorado Central's "new lead-zinc" mill that spring, calling it "another source of supply of lead and zinc." It gave the "famous old town" of Georgetown, which "in former days was a rich silver-mining camp," some needed national publicity. The detailed article covered the entire process that had been devised on "the results of a series of tests."[10]

Great publicity for a moment, but trouble appeared soon enough. The *Courier* reported on September 20, 1919, that "the flume of the Colorado Central mill broke last Saturday and let a flood of highly colored and strongly perfumed water into the town reservoir and water mains." Boilers and pipes throughout the town had to be drained "as best they could," the reporter laconically noted. Best was not good enough: "Many a one got a taste as well as a smell of flotation oil."

It broke again in late October, filling the "reservoir and service pipes with concentrations of water and oil." Then came the final act. On May 15, 1920, the *Courier* headlined "Colorado Central Shuts Down Temporarily." Apparently, the miners did not trust the company and on May 10, demanded to be paid their wages. The manager did not have the funds available and wired officers in the east for money. "They refused to honor the call."

The article continued: "So everything is in the air, and attachment proceedings were instituted. Manager George Collins blames himself for not suspending operations several months ago, when he realized that it was a losing proposition to operate the mill and dumps during the winter." The paper concluded, hoping "the eastern and western holders have been out of joint for a long time, but now that things have come they will work in harmony to save themselves."

During the rest of the rest of the month, the Colorado Central's troubles were the talk of Georgetown and Silver Plume. The eighty miners employed at the mine hired a lawyer and attached the mill, and the company truck, some personal property and forty tons of concentrates were sold to benefit the creditors.

It all worked out better than might have been expected, and, by the end of May, the men had received 80 percent of their wages and "anticipate receiving the rest before a great while." Other creditors filed liens, and the company went into receivership in June. The ever optimistic *Courier*, though, on June 23 quoted an unidentified authority who said "the Colorado Central Mines Co. will within a short time commence work in the mine and put it in first class condition level by level."

Dark clouds gathered over Leavenworth Mountain by late summer. The *Courier* had to report that "for the first time in 55 years no mining is being done on Leavenworth." Of all the properties on this once-famous mountain, only the Colorado Central "has been worked deeper than a few hundred feet." Fear not. The paper fearlessly forecast that "millions of dollars yet remain to be mined from near the surface." The Colorado Central would not be leading the charge, however. In November the paper reported that among the property sold at a delinquent tax sale was the Colorado Central.[11]

Sometimes that technique proved just an effort to freeze out some group or individual, but not in this case. Too often now mining news included such choice items as "the aerial tramway has been sold for junk"; "work has been discontinued on account of the lack of funds"; "the old compressor has been sold for junk"; the mill buildings "were sold to a wrecking company"; and finally, "the company has temporarily suspended work." The equipment came from closed mines and the others are self-explanatory.

Nor were news items such as the following indicative of a district in boom: "The mine is about to start operations and wants about six men." Small operations had come now to completely replace large-scale mining.

Not that there were not items to counterbalance such gloom. There most certainly were, and in greater number, as the *Courier* never forgot to mention each one with its usual enthusiasm and expectations.

> The Mid-Colorado Mining company has about twenty men employed in development work in the Mine and in overhauling the mill.[12]

> During the past year a very large amount of development work was done in the Capital Mine. This Mine has contributed largely to the output of ore for upper Clear Creek during the year 1917, and with the plans for future development now under consideration, will more than hold up its record this year.[13]

> The company is now planning to build a mill below the tunnel portal, which will be of the latest type.[14]

> The Georgetown Tunnel Company held its annual meeting here the latter part of the week and it is reported that they are planning to do big things at their holdings.[15]

> Thomas Teal, manager of the Pennsylvania mine near Argentine pass, was in town this week seeking miners and mill men. Mr. Teal is getting the mine in splendid shape for production and will operate the mill as long as the weather conditions will permit.[16]

It all demonstrated to the experienced mining analyst a district evolving into history and trying to keep its hopes alive.

Mining still held an allure to outside investors, who bankrolled many of these projects. The *Courier* illustrated this in a story about the Centennial Mine reopening: "It is understood that a long time lease and bond has been granted on the Centennial mine to Los Angeles and Eastern parties, and that a force of men will be put to work within a few days."[17]

They and/or their money arrived with great expectations, just as they had in 1864 and 1879, and generally left just as easily with broken dreams and a skinny wallet. Each new generation needed to learn anew that mining offers potentially huge returns, but equally large potential losses for its investors.

Meanwhile, depressed Georgetown and Silver Plume were evolving from strictly mining communities. They looked about to find another economic pillar, or pillars, which came, in part from their mining heritage. One possibility was tourism. The striking scenery of mountain-crested valleys proved alluring, even though they were marred by that very mining heritage. Regardless of what the old-timers who still clung to mining thought, changes were coming to the region.

The scars of a generation of mining in an around Georgetown were noticeable as the mine dumps, abandoned buildings, and rusting machinery once considered signs of progress and prosperity. In the midst of this environmental chaos, nature was starting to reclaim her own; eventually, it would be up to everyone to help the process along.

The *Courier* noted: "Not many years ago the western face of Griffith mountain south of town, was a barren zone, denuded of timber in the forest fire of 1867 [perhaps another impact of man]. Today there is a profusion of evergreens and aspens, which during the past week have taken on the colors of the rainbow."[18]

Silver Plume and Georgetown might be economically depressed, but that was better than their neighbors, the little mining camps that had once claimed notice here and there among the mountains. Bakerville, Waldorf, Lamartine, Chihuahua, and others of similar size, at one time or another rushed into the news with prospects for future growth and mining boom. Now, they decayed, forlorn relics of the past, ghost towns aborning. Only Empire appeared to have much of a future, thanks to the potential of molybdenum, but even it, at the moment, was languishing.

## PHOTO ESSAY 5

# TRAINS, TRAMS, AND TOURISTS

The Colorado and Southern railway wound its way from Golden to Silver Plume, where it joined with the Argentine Central that chugged along the steep hillsides up to the high elevation town of Waldorf. The Sunrise Peak Tramway, gave visitors to Georgetown and Silver Plume the type of thrilling adventures that they would try to capture with early cameras and write about in their diaries. The first railroad came to Georgetown in 1877 as a means of transportation for travelers and the ore produced by the mines. With the completion of the Devil's Gate viaduct—the High Bridge—in 1884, the railroad became a tourist magnet. By the turn of the century, tourism became increasingly important to the community as mining struggled. Once cars and trucks started to travel the roads between Denver and the mountains, however, rail traffic faltered. Passenger routes were discontinued in the 1920s, and the tracks pulled up in the late 1930s. Automobiles, traveling on newly widened highways, would become the wave of the future.

*Smoke billows above the houses in the south end of Georgetown as the train heads up to Silver Plume c. 1910. Ronald J. Neely collection.*

*The picture is taken looking east toward Silver Plume with the Colorado and Southern line in the foreground. The Pavilion and train can be seen in the center of the photo. Bradley collection.*

*A rotary plow in front of the engine blew snow off of the tracks between Silver Plume and Georgetown. The train is parked in front of the smaller Silver Plume depot, now used by History Colorado, c. 1910. Chappel family collection.*

*Tourists visit the Dibben's Mill or International Smelter, built in 1868 as silver mining developed in the East Argentine region. The stop may have been part of their ride along the Argentine Central from Silver Plume to Waldorf, c.1915. Denver Public Library, Western History Collection x1129.*

*Photographer Hook catches a blurry image as a train crosses the high bridge above Georgetown. The dam and original Devil's Gate bridge can be seen along with the penstock used for hydroelectric power on the main branch of Clear Creek. Ronald J. Neely collection.*

*The "Ice Palace" (actually a tunnel on the Johnson mining claim) was a popular stop along the Argentine Central railroad grade above Waldorf. Denver Public Library, Western History Collection, McClure photo MCC288.*

*The Argentine Central locomotive is parked above the buildings at the entrance to the "Ice Palace." Denver Public Library, Western History Collection, McClure photo MCC697.*

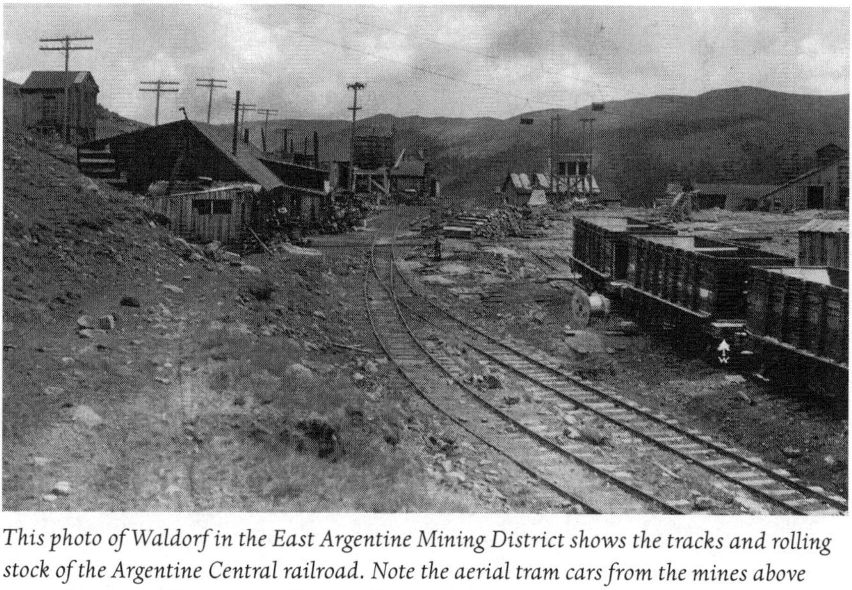

*This photo of Waldorf in the East Argentine Mining District shows the tracks and rolling stock of the Argentine Central railroad. Note the aerial tram cars from the mines above town. Denver Public Library, Western History Collection x123973.*

*From left to right: Catherine (Kate) Reilly and Ethyl B. Reilly, are shown standing in front of the post office in town of Waldorf c. 1910. Elliott family collection.*

*An Argentine Central train puffs away on the main street in Waldorf c. 1910. Elliott family collection.*

*As mining struggled, the tourist industry continued to grow—spurred by grand adventures such as the train across the Georgetown Loop, followed by the breathtaking Sunrise Peak aerial tramway built in 1906. Bradley collection.*

*A dance pavilion 20 × 40 feet and several small cabins were built along the Argentine Central line in 1916, two years before the line was discontinued. Visitors would continue to access the buildings by automobiles, or a good hike from either Georgetown or Silver Plume. The building became a destination for both tourists and locals. Art Davis photo, gift from Tom Young to Christine Bradley.*

*Constructed in 1906 at a reported cost of $26,480, the tram climbed over 2,000 feet of mountainous terrain south of Silver Plume. The sign announced the ride as "The World's Greatest Scenic Route." Art Davis photo, gift from Tom Young to Christine Bradley.*

*It is hard to imagine how the ladies' hats survived the trip up the Sunrise Peak Aerial Tram. Photo was a gift from Tom Young to Christine Bradley.*

*These buildings stood at the top of the Sunrise Peak Aerial Tramway. All that remains are a few towers on the hillside, which can be seen from a few locations in the valley floor. Chappel family collection.*

*Passenger travel along the Colorado and Southern ceased in the 1920s, though limited freight traffic continued until the line could no longer cover its costs. Consolidation Engine 65 heading toward Silver Plume in July 1938, possibly going to pick up rail from the abandoned line. Richard H. Kindig photo, Bradley collection.*

*A work crew struggles to pull up tracks in the vicinity of Alvarado cemetery (Empire junction), 1938. Marcellus Merrill photo, Marcellus Merrill collection.*

*Engine 68 is along to gather sections of track as they are pulled out in 1938. Marcellus Merrill photo, Marcellus Merrill collection.*

*The postcard speaks for itself, as the picture captures men beginning to take up rails and dismantle the High Bridge in March of 1939. Frank J. Duca postcard, "Courtesy of Buckley and Son" Bradley collection.*

*Engineers witnessing history as the old Colorado and Southern narrow gauge track disappears in 1938. Marcellus Merrill photo, Marcellus Merrill collection.*

# CHAPTER 6

## 1917

> When the bell-tower collapsed, I will never forget the sound of the church bell clanging all the way to the ground. It was like a death knell, for that beautiful tone had called us to worship ever since we were born, and our parents and grandparents before us.
>
> *Remembering the fire at the Catholic Church in "My Childhood In The House on the Hill in Georgetown, Colorado," Veronica M. Elliott*

The chill winds of January tucked an ember from the passing noon train into the wooden roof of Georgetown's Catholic Church rectory on Tuesday, January 2, 1917. The fire was detected immediately, with the assumption it could be easily controlled. Soon the wind blew harder, and water from chilled hydrants came out as a meager trickle. Before long the entire roof succumbed to flames, which quickly spread to the neighboring church structure. Hundreds of people stood in horror, some weeping, others praying, as the sturdy old landmark buildings burned to the ground.[1] The noon train had caused problems at the Clear Creek Mill at Empire Station and had started a small fire in the yard of the Henry Kneisel house on Biddle Street all on the same run. The Kneisel fire was successfully doused by Mrs. Kneisel. One passenger noted that the engine had spewed sparks throughout the day. The church buildings were insured for a total of $3,500, though claims against the Colorado and Southern would be settled for $15,000, which would be put to building a new church well clear of the railroad tracks the following year.

The proximity of the tracks to the church and concerns about fire had been acknowledged in 1885, leading the Methodist congregation to move their old wooden church, located just across the tracks from the Catholic church, downhill to the corner of Ninth and Taos Streets. The imposing Catholic church, built of brick, seemed indestructible and was far too heavy to be moved, staying in situ for forty years before succumbing to sparks from the train.

By 1917 the old Methodist church, while safe from fire, needed substantial repair. A grand plan was developed:

> The present frame will be straightened and firmly braced. The auditorium will be shortened to give room for larger Sunday School rooms in the west end of the building. There will be one room 20 x 27 which may also be used for a dining room or reading room when desired. A small kitchen and primary Sunday School room will be to the right of the present entrance. The two large rooms down stairs may be thrown open to accommodate large audiences. Where the old balcony is will be a large room 27 x 33, for Senior Sunday School and for social occasions. This room will also open as a balcony when needed. A new entrance on the south will greatly beautify the plant and will accommodate any who come for services before Sunday School is over.[2]

Fundraising efforts started with a series of lectures, continuing through the fall; however, the congregation could not complete the task during the hard times of the early twentieth century. The following spring the congregation sold the property to the Catholics, who were anxious to move away from the old location and the danger presented by passing trains.

The Catholic congregation also faced hard times and a dwindling membership. The church elders quickly decided to tear down the old Methodist church and build a small brick church, which could be paid almost entirely with the money from the damage suit against the railroad and pending insurance claims. The local newspaper editor scolded the church fathers for not digging into their own pockets; however, there were few if any wealthy parishioners still living, certainly not like those of the nineteenth century who had built the old church. The treasured bell, which had fallen from the steeple during the fire, was sold to a Los Angeles firm. The start of construction was delayed until 1918, as everyone felt that the prices would soon go down and allow them to save money.

One bright note came as the members of the Presbyterian church (who would soon gather additional members from the struggling Methodist congregation) managed to fix and replace broken windows in their church with funds from the "Christian Endeavors" group.[3]

The threat of fire plagued the town's aging wooden buildings, escalating when a structure at the corner of Sixth and Argentine streets succumbed to arson. In January 1917 the *Courier* noted the formation of a "vigilance committee"; however, no hangings were reported.[4]

Two months later in Silver Plume flames lit the night sky, spreading rumors of a fire at the Burleigh Mill or the Silver Plume School. The fire was confined to a small cabin in the vicinity of the Silver Plume School; however, the townspeople who showed up to fight the fire (the fire department having faded from existence) were unable to get into the nearby firehouse due to snowdrifts in front of the building. Once they had hoses in place, the hydrants were nearly useless since they were frozen, much like those at the Catholic church fire in Georgetown.[5]

Both towns struggled to keep active fire departments as they dwindled in size. The fire department in Silver Plume would regularly disband only to reorganize after a fire. Georgetown kept an active department due to its larger population but shared Silver Plume's difficulties. Both towns responded to their own and their neighbors' fires; however, the decrease in department membership (or total lack of a department in Silver Plume), problems with aging water systems, hoses, and equipment made fire fighting difficult at best.

Georgetown had reorganized its department in August 1916 under the leadership of new chief Miles Cunningham. By August 1917 the department reported fifteen meetings and 141 members with 135 also belonging to the state association.[6] The individual companies remained: Old Missouri, Hope Hose, Alpine Hose, and Star Hook and Ladder. Nevertheless, discussions were held regarding possible consolidation. While the companies struggled for new members, pride in the individual departments and their histories continued. The Alpine Hose Company sponsored a "smoker" in May 1917 with a dinner of "hot beef sandwiches, cheese, pickles, coffee, ice cream, cake, cigars and cigarettes . . . served in fine style, the tables being nicely decorated with the two great silver nozzles won by the company as prizes, as center pieces." The company members were entertained with stories from the past, including the tale in which the city council, which had used the building for town meetings, had been barred from the meeting room after the selectmen refused to stop spitting tobacco juice on the fine new carpet that had been donated by William Hamill.[7]

While ensconced in their newer "digs" at the Star Hook and Ladder building, the city council members did take actions regarding potential fire hazards, including the need to condemn buildings reported as fire hazards and to order the removal of those that had been damaged by fire but not yet torn down. In August 1917 the council voted to remove the old buildings along Sixth Street between Rose and Argentine. The town's commercial district continued to shrink, and many of the holes made by fires and razings would not be filled for years to come.

Sidewalks—always an issue—would start changing from wood and flagstone to cement. Mrs. Emma Johnson put in a cement walk at the corner of Fourth and Rose, which connected to the walk Mrs. Graham had put in the previous year. The town council asked the clerk to look into the town purchasing an entire carload of cement to be sold to those interested in putting in new walks.[8]

Amidst the loss of buildings on Sixth Street, Harry Hanson, manager of the hydroelectric plant, decided to clean up the lumber and trash piles around the building and planned to paint all of the buildings owned by the company. As might be expected, the paper encouraged others to follow suit: "This will make a wonderful improvement in that part of town. It is to be hoped that other companies and private owners of property will follow the lead of the power company and improve their property while improving the appearance of the town."[9]

In April a new Board of Selectmen and Police Judge J. J. White reviewed the state of the town and declared their interest in a "clean, better and bigger Georgetown," with hopes for new roads to attract additional tourists to help the struggling economy. Plans for the new and improved route across Loveland Pass had become the focal point as merchants in both Georgetown and Silver Plume lobbied local, state, and federal agencies. Early progress came to a halt when county attorney Reginnitter (of Idaho Springs) told the Board of County Commissioners that they could not expend funds within the municipal limits of the town of Georgetown even though the town had deeded the proposed section of highway to the county. Tempers flared, with the Georgetown editor noting that Mr. Reginnitter had not questioned the expenditure of funds for a bridge within Idaho Springs a few years previous. The "spitting contest" continued over the next several months, though it was noted that Georgetown and Silver Plume were getting along quite well: "Now that we are working in unison on one object, which is of as much importance to the state at large as it is to us, we must keep united on all measures and work to our mutual benefit, and not allow such things as baseball squabbles to interfere with our business welfare."[10]

Development of Loveland Pass as an automobile road became the project that everyone except an occasional county attorney loved, but no one could afford. The towns and county contributed funds and labor, but the amounts required overwhelmed their limited coffers. The county wrote letters, set priorities, and worked hard to find a combination of state and federal programs that might help. The Federal Highway Act of 1916 had created the Federal Aid Primary road system that began to fund projects of national importance. The Colorado Department of Highways began to expand their interests through these years, yet funds were limited. A blue ribbon panel traveled to Idaho Springs in 1917 to look at the needs and associated benefits of improving the road winding down Floyd Hill, which county residents hoped would encourage an interest in all of the major routes through the county.

State and federal agencies, however, would soon be overwhelmed by the costs associated with rerouting and completing improvements on Berthoud Pass, a well-established route needing approximately $225,000. Loveland Pass would follow the early route known as the High Line road, which was never as popular as Berthoud Pass to the north or Argentine Pass to the south. The new, improved road gradually inched west from Georgetown and Silver Plume but would not be completed until late in 1929. By this time the road was widely known as the Holy Cross Highway, running from Denver over to the area around the famed mountain site.

Named routes, whether built as circular day trips or parts of longer mainlines, became the norm, and the way to travel. The Denver Tourist Bureau realized the importance of promoting travel to points west of the city, dedicating $70,000, with no matching funds on the part of the mountain communities, to an annual campaign to promote travel to the mountains either by train or car.[11] Railroad companies did not yet view automobiles as a threat to their business, since most visitors still traveled to the state by train, then rented a car or hired a driver to access remote areas. Both Burlington

Northern and Union Pacific printed tourist brochures during these years to attract visitors to Colorado as part of joint railroad/automobile packages.

While the number of cars owned by locals steadily increased, the time required to travel from town to town was substantial. A group traveling from Denver to Idaho Springs in April 1917 planned to leave the State Capitol Building at 9:00 AM. and arrive in Idaho Springs in time for dinner. People stayed close to home, especially for entertainment, with the town movie theater providing a regular fare. One week in March, 1917 included the following:

> Saturday March 3rd—World. Carlyle Blackwell in The Shadow of a Doubt
>
> Sunday, March 4th—Metro. Mme. Petrova in The Scarlet Woman.
>
> Tuesday, March 6th—Metro. Ethel Barrymore in The Final Judgment
>
> Thursday, March 8th—Pathe. First 4 Episodes. Shielding Shadow and Luke Comedy, 9 reel. This program begins at 7:30 pm. 5 and 10 cents.[12]

Life at home began to change, as the newspaper noted that the "parlor" was becoming a thing of the past, now replaced by the living room: "The living room is all that its name suggests. Comfortably furnished, well lighted and with no useless ornaments to be sidestepped or knocked down, it belongs to the man just as much as to the women, which is, perhaps, the finest thing about it."[13] While the living room may have become the domain of the man, clubs and social groups remained the domain of the women in town. The Grand Opera Study Club and the Students' Literary Club met on a regular basis as well.

In January the Students' Club presented a program with a paper on the Lewis and Clark Expedition and a presentation on Guatemala. Twenty-five women attended the program, which included instrumental music by Mrs. Lendecke, Mrs. Pulsifer, and Mrs. Foley. The July meeting of the Grand Opera Study Club focused on *The Mikado* by Gilbert and Sullivan. Solos were sung by several members, with the remainder of the group acting as the chorus. Yokozawa San, a woman from Japan attending Colorado College, as well as Miss Ruth Regan, a former missionary, brought items for display and discussion.

The men in town were more interested in starting up an athletic club, though there were difficulties in finding a permanent home for the group and their equipment. Schoolboys placed a basketball hoop in the side yard of the old brick schoolhouse.[14] Fraternal organizations such as the Masons and

Oddfellows, along with the various companies of the volunteer fire departments, gave the town's young men something to do, combining volunteer efforts with social gatherings and entertainment.

The census of school age children from 1917 showed 189 children between the ages of six and twenty-one in Georgetown. Education was compulsory up to age sixteen or graduation from eighth grade, which placed the emphasis on grade school education and kept the first-floor school rooms at the old school quite full, especially with the closure of the Catholic school. The Georgetown school district provided a four-year high school program for interested students, though class size dropped as some children decided not to continue beyond eighth grade. The detailed annual report filed by the district secretary the following year would show 180 of that year's 194 eligible students enrolled, with 147 in grade school with an average class size of 18 and 33 in high school with an average class size of 8.[15]

The graduating class of 1917 included four boys: John Frederick Keating, Frederick Lawrence Nelson, Robert Manley Nelson and Tom Moore Straub, as well as three girls: Balel Levinia Poirson, Arveda Katherine Swanson and Elva Violet Williams. The class colors were blue and gold, the class flower yellow rose, and the class motto "Finis coronat opus," or "The end crowns the work."

The excitement surrounding graduation and the ceremonies at the Opera House / Gymnasium were intermingled with fears related to the war in Europe. President Wilson had announced a national draft requiring all men between ages twenty-one and thirty to register. The school program included the song "Just Before the Battle Mother" about a mother at home and her soldier boy at war.[16] Mrs. Burkholder planted potatoes in the lots to the west of the Hotel De Paris as part of the national victory garden movement.[17] The young men in town decided to practice military drills in case they were called to war. In April it had been announced that "two secret service men have been in the county investigating the pro-German activities of a prominent citizen of this county."[18]

The new principal of Georgetown's school referenced German farmers in his report aimed at encouraging higher standards for American students:

> We oftentimes wonder why the German farmer secures a greater profit from ten acres than the American farmer from his one-hundred-ten. The

former plows deep, conserves the resources, plants between the rows, farms intensively. The latter scratches the surface, dissipates resources, trusts to Providence and the climate to give the increase, and gets the name of farming intensively. Just so with the school. It may have enormous numbers, large graduating classes, and yet the instruction may be so extensive, superficial, non-practical that the students in life find themselves so little equipped for the keen competition they must meet in the outside world.

I shall enumerate some of the changes inaugurated in our school during my regime, that you may judge whether progressive and in keeping with the times.

There is now a regular system of writing pursued at a regular period. Not a ten-cents worth of foolscap paper and write when you please. Likewise we have drawing pursued in the grades and high-school. In nearly every vocation or study in life drawing is one of the essential features. Can you recall when you had rudiments of music taught in all grades in the school? Last spring I requested the board to secure a teacher who would teach music in the grades. They did better, be securing teachers who can teach music in their own grades.

In the high school have been taught the rudiments, Cooke's Music History, and many lessons given from Okem's harmony treatise. The grade school teachers by instructing the pupils after school hours have made it possible for a much greater number to make their grades this year than otherwise would have done.[19]

The stern tone of the principal's letter reflected the fears and frustrations of a nation at war. The townspeople kept their hopes up with patriotic songs and other actions. The town council passed a resolution requiring flags to be flown at each fire station. The Alpine Hose and Old Missouri flag poles had to be removed and replaced, and a pole was erected at the Hope Hose Company building since none had ever been built. The paper noted that the cost was substantial, but "The town now looks more like it ought to in war times."[20] The women of the Students' Club joined in with a program on the Star Spangled Banner and Flag Etiquette by Mrs. Hughes, concluding with the singing of national songs.[21]

The world at war and the need to supply troops with food and armaments created shortages and a need to plant gardens to grow food for individual families. The town celebrated National Food Day on July 1, 1917. The *Courier* announced the upcoming event:

> Under date of June 18, Herbert Hoover mailed to every minister whose name he could secure a personal letter asking him to preach upon the subject of Food Conservation on Sunday, July 1. He expressed the hope that this day will become known as "Food Saving Day," and that it may be made the purpose and motto of every household in America to Save the Waste and Win the War. Mr. Hoover urges the enlistment of civic, fraternal, patriotic, and other organizations in the effort to make Food Saving Day really effective in conserving the food supply for us and for our Allies, and in reducing the cost of living to our own people.[22]

Some suggested plowing the city park to plant vegetables as part of the national effort.[23]

The Parent Teacher Association sponsored a municipal exhibit at the Red Cross Hall in the fall, where local gardeners brought forth a variety of fruits and vegetables including apples, crabapples, strawberries, cabbage, carrots, turnips, rutabagas, potatoes and a kohlrabi grown by William Reichwein.[24]

Georgetown and county Red Cross chapters were established in June 1917. In their honor, Mayor White welcomed the crowd to the Georgetown opera house and made note of the three Georgetown boys already serving in the armed services: Leonard Katzemeyer, Raymond Tong, and Ned Burkholder. The Red Cross hoped to provide funds for the care and well-being of the troops as well as giving those not able to serve the opportunity to make things ranging from bandages to candies and sweets. Mrs. Burkholder encouraged the ladies of the chapter to join her in the Fourth of July parade by dressing in white and wearing their Red Cross badges. A date was planned for instruction in making of bandages and other sewing work. Fundraising started shortly thereafter with impressive results. Girls from Silver Plume gathered $400 in two days.

The first group of draftees headed out of town in early October, with hundreds of citizens present. Schools had been dismissed for the day, and all gathered at the Georgetown depot to salute the boys. The schoolchildren sang an emotional rendition of "America" and waved as the county's finest pulled out of the station: Ralph Wilkins, A. F. Arthur Fallgreen, Alva A. Ross, Robert W. Callow, and Crist P. Staingier of Idaho Springs; William G. Harper and John N. Walder of Alice; John M. Howard of Lawson; Emmons J. Hunt of Dumont; James J. Varolotti of Silver Plume; and Andrew V. Bartolotti and Thomas M. Silver of Georgetown.[25]

Amidst the fears related to war and local boys heading overseas, the need for materials soon created an expanded market for local ores. By the middle of 1917 the government started making specific requests relating to molybdenum, vanadium, uranium, nickel, manganese, platinum, and large pieces of transparent unfractured fluorspar. An article in the *Courier* encouraged those with information related to potential deposits to contact the state geologist in Boulder with samples when possible.[26]

Parents and children walked in fear as news of battles in Europe and casualties began to filter back to the quiet confines of Georgetown; however, in many ways the dangers of mining had a greater impact. The Capital Prize Mine at the east side of town had a particularly bad year. In January, 1917 Nels Nelson from Denmark was struck on the head by a rock. In June Fred Bellotti was killed by rock fall as well. In December Louis Santer was badly hurt, and later that month J. D. Iverson was killed after he lost his balance and fell as a result of bad air. That same year twenty-year-old Harry Morris was electrocuted at the Imperial Mines Mill near Waldorf. The deaths from mining in 1917 would outnumber the town's death toll from the war in Europe.

Everyone knew mining was a dangerous occupation, subjecting its practitioners to falls, blasts, rockslides, bad air, and deadly diseases such as silicosis. The decision to send troops into battle in Europe and the subsequent draft put the country's young men in danger, yet by the end of 1917 the deadliest threat lay ahead and would claim more lives than local mines and the war to end all wars. Quietly and without warning, the world began to feel the effects of pandemic flu.

# PHOTO ESSAY 6

## GEORGETOWN, 1917–1929

*The interior of Our Lady of Lourdes Church on Main Street is shown shortly before the fire in 1917. The building was completely destroyed, but the congregation and helpful townspeople were able to save the stations of the cross and some of the statuary. Elliott family collection.*

*The interior of new Catholic church on Taos Street (c. 1920) shows the statuary saved from the fire on either side of a much smaller altar. Elliott family collection.*

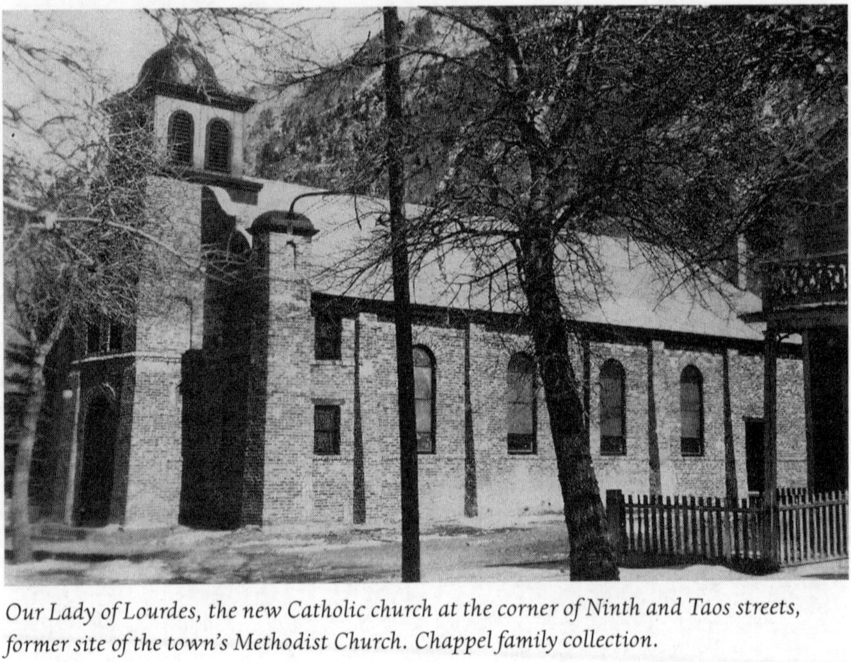

*Our Lady of Lourdes, the new Catholic church at the corner of Ninth and Taos streets, former site of the town's Methodist Church. Chappel family collection.*

*The front view of the new church shows a lovely brick building, even if the local newspaper editor felt the congregation should have done more fundraising and built a larger structure. The building still stands as one of a handful of buildings built during the hard times between World War I and the depression. The photo may have been taken after the funeral of Father Walsh. Elliott family collection.*

*Young men headed for active duty as part of the 1918 draft stand in front of the depot in Georgetown. From left to right: Herman Crist, Clarence Gunstrom, James Florine, Otto Pilz, Manning Barber, and [probably Frederick] Leach. Bradley collection.*

*Ned Burkholder was one of the first Georgetown men to volunteer for service in World War I. Hotel De Paris collection, Colonial Dames of America, Colorado chapter, Burkholder collection.*

*The "Floyd Hill Auto Drive" was the main route from Denver to the central Rockies for many years. The trusty Model As and Model Ts traveled routes many would now call four-wheel-drive roads. E. J. Deis postcard, Bradley Collection.*

*Counties and states struggled with the need to make roads wide enough for two cars as the amount of traffic burgeoned. This caravan of "tin lizzies" is just east of Dumont traveling toward Idaho Springs probably about 1920. Image courtesy of History Colorado (Marchington collection 10033982).*

*The road west of Silver Plume served horses and cars alike. Image courtesy of History Colorado (Neal Fisk collection 10032939).*

*The Coggshall family and friends driving down a canyon road c. 1915. Coggshall family collection.*

*As the demand for electric power increased, the need for transformer houses and a network of dams south of Georgetown became apparent. This transformer house stood at the southeast end of Sixth Street next to the hydro-plant, 1920. James G. White photo 5140, Charlie Wright collection, John Tomay Memorial Library.*

*The Ft. Collins flag makes it seem as if the trip may include a football game. c. 1920s. Hotel De Paris collection, Colonial Dames of America, Colorado chapter, Burkholder collection.*

*Old town and new show up in this Glenn Gebhardt photo of Sixth Street taken in 1926. Photo courtesy of John Tomay Memorial Library.*

*It is hard to know if the young ladies are models or campers with their tent set up on Berthoud Pass c. 1925. The state of Colorado was anxious to attract visitors to its mountain roads. Georgetown merchants tried to make sure they were represented in state publicity and advertising. Denver Public Library, Western History Collection z7609.*

*Not all of the old mining roads were suitable for auto traffic since horses with wagons could travel steeper grades. The road up Stevens Gulch managed to handle both types of traffic. The network of county roads changed dramatically during this time. Image courtesy of History Colorado (James Grafton Rogers collection 10039781).*

*Automobile campers serve up some grub in the valley west of Silver Plume. The automobile and improvements to area roads made access to the wilderness easier. Image courtesy of History Colorado (James Grafton Rogers collection 10039769).*

*A hiker stops for a rest west of Silver Plume c. 1920. The four lanes of pavement that came with the later construction of Interstate 70 couldn't have been imagined at this time. Image courtesy of History Colorado (James Grafton Rogers collection 10039770).*

*James Grafton Rogers and friends attempt a few car repairs in the days before AAA. The sign points the way to old Berthoud Pass route west of Empire. Image courtesy of History Colorado (James Grafton Rogers collection 1039772).*

CHAPTER 7

# THEY ROARED AND SPUTTERED

## *Georgetown Moves into the 1920s*

The War to End All Wars ground to a halt. The world now seemed safe for democracy and Christianity with the barbaric "Hun" defeated. The boys would soon be coming home, and peace would return to the land. Georgetown, like other Colorado towns, celebrated that eleventh hour of the eleventh day of the eleventh month of 1918, when the killing stopped. Enough of war—people awaited the blessings of peace, but first their goal, their joy, was to celebrate victory, with Georgetown not missing a beat that day.

Schools were dismissed, stores closed, dynamite "booms" echoed around the mountains, and in the afternoon "pupils assembled in the business section and all joined in singing the national hymn." The rest of the day, and into the night, was spent "ringing bells and making joyful noise generally."

Folks celebrated all right, but something else, even more deadly, stalked across the world, the Spanish Influenza. Before it finally ran its course, some 30 or more million people would be dead, including 7,783 Coloradans, who died during the ten months of the worldwide pandemic. The flu proved especially deadly at higher elevations and among young adults. Silverton, for instance, lost nearly 10 percent of its population, a higher percentage than any other town in the United States.

Fortunately, Georgetown came nowhere near to that unwanted and unwelcome claim to fame. It could not, however, escape this virulent plague. The *Courier* sadly documented its tragic course during the peak months of October, November, and December, 1918. Starting on October 12, the paper carried a column "Victims of the Spanish Influenza." That issue also gave advice from the Surgeon General that the flu "is a very contagious kind of 'cold' accompanied by fever, pains in the head, eyes, back or other parts of the body." It "ordinarily lasted three to four days."

If one became sick, the important thing to do, the Surgeon General recommended, "is to go home at once and go to bed." That would keep the patient from scattering the disease "far and wide" and also "help keep away dangerous complications."

How could one keep the influenza away? Avoid "overcrowded homes" and all public gatherings, dress warmly, and get proper "proportions of work, play and rest." It also helped to eat "sufficient wholesome and properly selected food." Actually, nothing helped. No known medicine combated the illness successfully, and doctors offered no miracle ideas; consequently the epidemic only just grew worse.

Quarantine signs placed on homes with flu victims warned one and all: it "is folly to visit a house with a quarantine posted sign." Theaters and schools were closed, public meetings canceled, and even church services forbidden. People were to stay home as much as possible and not mix in crowds, and they were to take "all precautions possible" and avoid "carelessness."

Some people begrudged the fact that prohibition was the law of Colorado, thus a shot of whiskey could not be legally taken. Alas, there went that favorite old-home remedy. The flu proved particularly deadly when it came in combination with pneumonia. A person could come down with the flu in the morning and be dead by evening, and there seemed to be nowhere to turn to avoid it.

Two weeks later, the *Courier* gloomily noted that there had been about 200 cases and thirteen deaths in Georgetown. Short obituaries were written about of the victims of this "terrible scourge" and more hopefully, news that the sick "are recovering." Confirming the general trend, the article observed that almost all the victims were youngsters or people in their twenties and thirties.

The flu continued on, peaked, seemingly slowed, and then on November 30 the *Courier* warned its readers: "The influenza, which was thought to be under control in this county has taken a new start at Silver Plume and Idaho Springs and the lid has been again clamped on at both places." The paper diligently published lists of the ill along with the deaths in Silver Plume and Idaho Springs.

By mid-December, the flu waned and even left the front page of the paper. The bans were lifted, quarantine cards almost all taken down, and "flu patients recovering" was the good news. Only a few cases now were reported, and the worse seemed over. Although in the spring the flu returned, it was fortunately in a milder form.

In the midst of all this, the *Courier* reported that John Dumont, "probably the only living pioneer of 1858," had died in Denver. He had known the Griffiths and other pioneers and once owned mines at Empire and near Idaho Springs. The ties to those long-ago days were gone. The new era of the roaring twenties, prohibition, and the flapper slipped into Georgetown, Silver Plume, and the rest of Clear Creek County, sometimes with a roar, sometimes almost silently.[1]

The 1920s dawned—the unpredictable 1920s, the decade of letting a nation's hair down after the reform years before the war and the all-out effort to defeat the Germans and his hordes during the war. Seldom has America seen such social and general change. Before the decade was over, radio, talking motion pictures, and airplanes would become commonplace. The automobile, credit buying, and a host of things for the home and individual had changed life and the times forever. Women now could vote, smoke in public, go to a speakeasy, joyride with men in their "merry Oldsmobile," and wear clothes that would have shocked their grandmothers—and perhaps their mothers.

The big bull market, with its glowing stock reports, elbowed other news aside, and people speculated in stock who had never ventured into this world before. They also bought appliances, homes, and cars in installments. No longer need one deny him or herself something by working to save money until it could be purchased; the old-time virtues went out the window. Behind all this lurked the other side of this new world.

People wondered what was happening. Prohibition had finally become the law of the land, after decades of work by the Women's Christian

Temperance Union (WCTU) and other groups. People broke the law with pleasure and enjoyed doing it, something unheard of in earlier years. The fight over Darwinism and evolution stirred some heated emotions, while others joined the Ku Klux Klan to preserve the "old values." A few discussed what the "Red Scare" might be about; did communism and socialism really threaten America? Youths enjoyed joyriding in those cars and doing what their parents dared not imagine. They also flocked to movies that promised "brilliant men, beautiful jazz babies, champagne baths, midnight revels, petting parties in the purple dawn."[2] Where was America headed?

Against this background, Georgetown "the Silver Queen of the Rockies," slowly was being transformed. Its best mining days had receded into the past, though a few hardy souls still held out dreams and searched to find someone to grubstake them and their mines. *The Mountain States Mineral Age* (August 1922) did not give up hope for the "mining capitol of the famous quadrangle." Mining was not dead, "but sleepeth. War conditions, followed by inflation, profiteering, oil booms etc., have each had their share in effecting the stagnation of the mining industry." Other businesses were affected too, "but not near so badly."

Georgetown of the 1920s looked increasingly like the rest of the country than ever before in its history. Its isolation hurt it more than ever before and, instead of better train connections, the town needed better roads to the outside world.

Tourism presented still greater possibilities, perhaps the best that were available to the townspeople. Mining could be a significant factor in this hope. It happened in a variety of ways. Mining man William Hamill's gracious home, one of Georgetown's finest, was converted into a "popular tourist hotel." Meanwhile, in nearby Idaho Springs, a three-day celebration honored the "Days of '59." The 1922 event tried to recapture those exciting days of yesteryear, with what have to be considered some strange (and today unacceptable) events—an "old fashioned miner's house burning, duel, lynching and hanging scenes, and turkey chases." These "Days" also presented some events that evoked more of the spirit of the age—placer mining, rock-drilling contest, gold panning, and an "air drilling" exhibition at the Newhouse (Argo) tunnel.[3]

Georgetown's once lively and promising neighbors of not so long ago had slowly become "ghost towns." They lured increasing numbers of sightseers, who came to wander about and wonder at what transpired there. A few ventured to abandoned mines and speculated about the silver fortunes taken from their depths, while searching over the dumps for a piece of glittering ore as a souvenir. Mining emerged in a different way as a tourist attraction and continued as an economic pillar of Clear Creek County. One could easily argue that tourism offered more of an economic pillar and held out a brighter future.

This was shown in the paper, as fewer mining notes or stories appeared. With the coming of spring, a few optimistic reports appeared, as prospectors and miners went into the mountains. Follow-up stories of successes proved as scarce as dandelions in a Georgetown January.

Tourism, however, grabbed the newspaper's attention. It encouraged readers to capture the tourist dollar, as these 1927 comments illustrate. "How about that cottage you may have for rent a little latter on? Just a bit early perhaps, but a small improvement here and there may make renting it much easier when the time comes." As summer neared the encouragement continued: "You might fix up a spare room and bring in a little revenue." Furthermore, "be sure to list them with the publicity bureau."[4]

Tourists came by auto now, and Georgetown aimed to collar them. The *Courier* hailed as a "commendable project" that same year the need to place roads signs directing drivers to the community. Money had been raised to paint and place a sign at the junction of the Berthoud Pass and Georgetown roads, "directing travel to Georgetown and Silver Plume." This sign was badly needed, "as motorists have been known to travel as far as Hot Sulphur Springs looking for Georgetown. The stranger motorist depends greatly on road signs to guide him and placing of these signs will bring results."[5]

The Georgetown Board of Mines and Commerce also did all it could. It encouraged people to write for information on "mining, business, and tourist opportunities." Of the three, the best hope rested with tourism. The *Georgetown Courier* published historical articles about the town's "exciting and fascinating past."

Georgetown stood ready. It offered a variety of things to do and things to see if only the tourists would come.

That aside, Georgetown found itself caught up in other issues. Despite holding out such great expectations to produce sweeping personal, moral, family, and community reform, prohibition failed to deliver such windfall blessings. The "promised land" failed to appear in Georgetown or elsewhere. Indeed, almost the opposite happened.

As early as 1921, the *Courier* pointed its finger at that sinful city of Denver, where "automobiles and booze have been great influences in corrupting youth." The two went together, no doubt. Old-timers and parents heard about the "petting" going on in those cars on dark rural lanes, or lonely mining roads, and some were moved to call them mobile "houses of prostitution."

To many people, however, "bath tub gin," and its associated friends, posed more of an immediate threat. The number of bootlegging cases in the county court increased steadily during the decade. Articles appeared about the sheriff seizing stills; one at Dumont, in 1925, had a seventy-five-gallon capacity. That, the paper noted, "was one of the largest stills ever found in the county."

Clear Creek County folks showed their true feelings when the "wet amendment," to allow the sale of liquor in the state, passed locally but not statewide. Meanwhile, the district court docket continued to feature cases involving "violation of the prohibition law." Despite this, the *Courier* told its readers where it stood. "Much bootleg liquor is poison and many deaths result, but after all, there is little danger of bad liquor killing good men."[6] Danger did lurk "in demon rum," as the prohibitionists used to say. The number of people killed or blinded or in some other way impacted by bad bootleg liquor should have given everyone second thoughts about indulging.

It was not hard to find liquor if a person possessed that desire—or had an overwhelming thirst. Booze, in all its forms, could be purchased locally. One had only to know the right person or persons. Somewhere in a nearby community, if not Georgetown, a speakeasy operated, perhaps using an ice-cream parlor or some other business as a front. No person need be denied the pleasure of having a drink combined with the added thrill and the excitement of breaking the law. People who would never have thought about lawbreaking before did it, even on social occasions at home.

Another unexpected change came about. Women, who would have been horrified at the thought of going into a saloon for a drink, and thus ruining their reputations, now freely ventured into speakeasies. That proved a tiny part of a larger movement, women moving beyond the restrictive Victorian boundaries that had been established for them so many decades ago. People noticed this in their dress, attitudes, job opportunities, social activities, community involvement, and increased participation in politics. They gained the local, state, and national vote. Particularly, the young seemed to be rushing beyond where their mothers had walked.

For many women, maybe most of Georgetown's, this new world did not impact them quite that much. They still preferred the lifestyle of their younger days and the role of women then. For them, being a wife, neighbor, homemaker, and perhaps a mother, as well as getting involved in the church, PTA, and maybe a club, provided enough activity for their days and weeks. Some worked in a store with their husbands, or in one of the towns businesses; they had little extra time for the flapper image no matter how popular it might be.

Depending on the family income, the world of the 1920s could bring relief to many women from the drudgery of housework. Electricity had promised and finally produced wonders to aid all aspects of work in the home: the refrigerator, vacuum cleaner, electric stove, and washing machine, to name just a few of the more obvious ones. Previous generations would have loved to have had these time-saving, work-easing, and modern marvels.

The radio, too, brought change to Georgetown. This, the *Courier* thought, provided a definite benefit. As its October 24, 1924, editorial sermonized: "We are indeed living in a radio age. It can truthfully be said that this has become a radio world. As a home builder and conservator of family life, radio has come at an opportune time—millions of families gathering in the home to hear the news and enjoy the music. From daily market reports for the farmers, baseball scores for the boys, religious services for the old folks and the jazz music. For the young, the radio is the age wonder of humanity."

All was not positive, however. When the governing board of Denver University refused to allow the November 1927 DU versus CU Thanksgiving day football game to be broadcast, for fear of a smaller attendance, Georgetown fans rose up in arms about losing this popular contest. They had "nothing complimentary to say" about such a development, but they could do nothing. Welcome to the new world!

Meanwhile, regarding that 1920s concern about the family structure changing, though not a particularly new one, many now thought they knew what the primary cause was. The automobile, as mentioned, allowed the youth more freedom and opportunity for activities outside the home than ever before, and movies showed them the up-to-date "modern" lifestyle. That worried the more conservative people, as well as parents. They looked about and saw the old ways and standards, the hallowed "good old days" vanishing amid the roar of the 1920s.

Their fears were not allayed when their hero, the silver-tongued orator, William Jennings Bryan, died in Dayton, Tennessee, at the conclusion of the sensational Scopes's trial regarding the teaching of evolution. Bryan and his supporters won the skirmish but lost the battle. He had planned to launch "another crusade against modernism" within weeks. Its success, though, seemed highly improbable, as urbanites shook their heads over rural "hicks." Rural America longingly looked back to an imagined "golden" agricultural past, while urban America looked to an imagined exciting urban future.

That "old-time religion" was not selling well among Georgetown's youth either; the secular world of the twenties marked the end of mid-week and Sunday evening services for many congregations. Too many nonreligious activities and forms of entertainment tugged at members' time, and the automobile, with Sunday drives, did not help either. Nor did the "the picture show theater" with its Wednesday features. Like their contemporaries, Georgetown's churches found their role diminished within the community, continuing the trend that dated back to the nineteenth century. The 1920s were not a good time for churches, apart from the faithful members who managed to keep their congregations alive and functioning, despite the trend of the times. The church appeared out-of-step.

Another group concerned about social cultural change slithered into the decade, the Ku Klux Klan. There can be little doubt, that there were Klan members, and sympathizers, in Georgetown during its mid-1920s peak, but as far as is known no Klavern, or local chapter, existed in the community despite Colorado being one of the two strongest western KKK states.

The Ku Klux Klan twisted Colorado in the mid-1920s, as its supporters elected members as governor and US senator as well as holding other state and local political offices. Why might a Georgetown individual join the Klan? It might have been over a genuine concern about the upbeat 1920s and its morality, or they might have enjoyed the mumbo jumbo, secret rites, and feeling important as a member. Small-town America also felt it was being left out of "modern" America, and some people in such communities believed they had been pushed away from mainstream America. Sadly, it was often because of personal hatred toward minority people. The KKK offered "solutions" for all these worries.

It would seem logical that Georgetown members would be associated with the Denver Klan, the state's largest and strongest, where rallies and cross burnings aroused the faithful into a frenzy of "anti" feelings. The capital city was also the headquarters for the state Klan. The Georgetown newspaper maintained a discreet silence about the KKK.

One of the ongoing sagas of the 1920s involved the completion of an automobile road over Loveland Pass. Georgetownites waited with bated breath for news about this issue, because it promised to open up the Western Slope for tourism right through their town. Wait they did.

In 1921, encouraging word surfaced that the State Highway Commission had decided to build the Loveland Pass road to replace the trail that went over the pass. The *Courier* soared with expectation. It "will cut time to the Western Slope to one day," compared to the train's overnight trip. Best of all, it will be the "shortest, easiest, most scenic route. It will become the most popular."

And they waited, as time drifted away. First, it took until January 1926 to complete the final surveys to the eastern crest of the divide and prepare the plans. The state intended to start "with construction work done by convict labor next summer." The prisoners labored "most of the summer above Silver Plume," until cold weather stopped them in November. The plan was to reach the top the following summer when the Western Slope road, with its "easy grade," would take only a "small amount of new work." The *Courier* continued to be optimistic about this "important link in the east and west highway chain. It will be a great asset when completed."

Indeed, it would. Yet two more years slipped by, and the road was not finished. The 1929 highway budget gave the project $30,000, "sufficient it is said to complete the road on the eastern side and put it over the top and part way down the western side." It never happened as planned. That October, Georgetown and Silver Plume men "went over the top of Loveland Pass," and a "goodly force of workmen cleaned up the old wagon road down the western side." Work hard they did, stopping for a picnic at noon, and made the road "passable but rough for a distance of a couple of miles where a big mud hole was met with."

As the 1929 panic settled into the 1930s Depression, only nine miles remained on the Western Slope of the still uncompleted pass road.[7] Locals had lived on hope for the past decade and, despite progress, still awaited that connection. Now, however, with times so depressed, the promises of the 1920s seemed lost.

Even with the delay in completing the pass, the automobile changed Georgetown forever, but no more so than by ending the town's marriage to that nineteenth-century wonder, the railroad. People did not ride the slow, time-consuming, and infrequently scheduled trains much anymore, or ship freight, for the same reasons. The Colorado and Southern Railroad responded, as other Colorado railroads had done, initially by curtailing schedules then asking for abandonment.

The first public hints regarding possible abandonment appeared in 1927, which aroused the Clear Creek County Metal Mining Association to "fight to have passenger service revived." They and others feared the freight service would be next, and right they were. Despite protests, the Colorado Public Utilities Commission, initially in 1927, granted the railroad's application to "abandon experimentally for a year all passenger traffic." This included that tourist wonder, the Georgetown Loop. Then the commission denied a rehearing, after deciding to allow the abandonment of passenger, but not the freight, business. In March 1929, the Silver Plume Depot closed, because the freight business did "not pay the agent's expenses."

The *Courier* warned its readers that "not only the people of Silver Plume but the people on the entire line might well take warning at this action." The editor predicted that this was but the "forerunner of what will eventually come to each station on the road until freight is stopped."[8]

This, he concluded, should not be blamed solely on the Colorado and Southern. If the people would not support the railroad, "it will," he wrote, "be the fault of the people, who in their own greed allowed the trucking companies to haul the freight which rightfully belongs to the railroad. Let this be a warning to Georgetown." There would be little it or its residents could do. This link to their past would eventually be gone also.[9]

Another cause the paper took up was the trend of people going outside of their community to do business. Georgetown was not alone; similar trends and editorials appeared in many other communities and had since the first mail order catalog arrived in post office boxes forty years ago.

As the decade closed, with hard times already on the scene, the paper rose to defend its merchants. Article after article, in the last months of 1929, carried the theme. Headlines such as these left nothing to the readers' imagination:

"HOME TOWN TRADE IS VITAL TO A TOWN, AIDING MERCHANTS ADVANCES THE TOWN, and SPEND YOUR MONEY IN GEORGETOWN, YOUR HOME." What would this do? It would retain prosperity, allow the town to grow, and support the merchants because "no group is a bigger booster and supporter of schools, etc." It did no good. Local business could not compete in price, nor variety, of merchandise available. To make matters worse, their customers faced hard times themselves and often could not pay their monthly bills, or even do much shopping. Their late nineteenth-century counterparts would have completely understood their twentieth-century plight. Not only was Georgetown losing business, but it was losing people. Initially this appeared somewhat newsworthy in a positive way. Marriage took some young women away to California, which apparently beckoned alluringly.

Finally, the *Courier*, took editorial action to rally those left behind:

> Just as in every town or small city, many a man in Georgetown gets the idea he must go elsewhere to make a fortune or to become successful in any degree. Distant fields always look richer, deceiving so even to the man who had to dig beneath his feet to unearth untold treasurers.
>
> The best advice that can be given to the average American is to cut out the wanderlust and apply himself at home.[10]

The paper repeatedly returned to this theme as the decade passed, while the editor continually tried to convince his readers of the joys of small town America: "In a town the size of Georgetown is where one really enjoys a house. In a city one is born in a hospital, reared in a boarding school, educated in college, courted in automobiles, married in a church . . . spend mornings golfing, afternoons playing bridge. Live in Georgetown and enjoy your home." Such thoughts did not help much. The trend rolled against such optimism. For example, both the county and the town consolidated voting precincts in 1922 to save "quite a sum of money." Georgetown found itself with only one, because of "its loss of population." The census returns of 1920 and 1930 further illustrated that same fact. The population declined dramatically from 703 in1920 to less than half that ten years later, with only 303 residents remaining, an alarming trend.

While many individuals advanced ideas, or solutions, about what might be done concerning this discouraging drift, the answer was elusive. Georgetown correctly foresaw that the best chance for its future lay in its past and tourism.

However, preserving that past ran up against a shabby-looking present, and the latter won out. The *Courier* of May 15, 1926, editorialized about buildings being torn down. Some comment, the editor noted, was occasionally heard "criticizing the tearing down of so many buildings." He sympathized. Having lived long enough to "remember when there were no vacant buildings in town," and it had so many more buildings than currently, it was "no wonder that it causes a heartache."

However, the editorial went on to express the view that the buildings going down were not fit for occupancy. "The general opinion is that that the town looks more prosperous and more sprightly with vacant lots than a lot of tumbled down buildings giving the streets a forlorn, deserted appearance. These buildings are valueless and fire traps that can not be used. If torn down that would add greatly to the appearance of the town." Another heritage of the past disappeared in the name of civic improvement. The conflict between preservation of nineteenth-century Georgetown and modernization snuck into town. The issue would only grow as the year passed.

The old days did not completely disappear. Georgetown's baseball team still took to the diamond to uphold the community's reputation. That is, it did most of the time. In the early 1920s, in some seasons, only the high school fielded a team. Part of the problem may have been with the town board of selectmen. In 1922, it decided to convert the "old ball ground" into a camping place for tourists. Mighty Casey and his whole team had struck out to the tourists' needs of the roaring twenties.

Lack of a town team moved the *Courier* to encourage the "boys" in 1926. "We haven't had a team in two years and there is no reason to let another year slip by. Baseball is one of the cleanest and best of sport and material is at hand for an excellent organization and it deserves the hearty support of every loyal citizen."[11]

Another year did not "slip away," and, for the rest of the decade, the Georgetown nine took to the diamond, once they found missing uniforms: "Suits cost considerable money and are valueless to anyone but a ball player and are necessary to the ball player." When the team won three games over the Fourth. the paper crowed, "it showed what practice can do." Throughout the season, the paper and fans cheered their boys on, and, when they defeated Idaho Springs in September 8 to 4, it was declared the "finest and fastest game of the season and close until the last few innings."

Baseball also gave the locals a chance to gloat, particularly when a Denver nine went down to defeat. That occurred in August 1927. "Of course the teams from the big city come to the small town with the idea that they are just going to make a killing. This feeling was strong in the mind of the visitors last Sunday, but they returned home that evening a wiser bunch." Then came the pièce de résistance, "at no point in the game did they stand a chance of winning." Georgetown scored a "fine victory," 11 to 3.[12]

Victory, though, did not always shine on the Georgetown boys. In June 1929, they went to Erie and "got a hard walloping," 16 to 7. According to the story, "hard luck" and the "absence of some of the best players" accounted for this result. Grudgingly, it was admitted, Erie "was a little too much for them."[13]

The Fourth remained the biggest holiday of the year, featuring, as it had for years, fireworks, youth games, "all kinds of races," a dance, and the traditional baseball game. Thus it was, in 1928, when a doubleheader highlighted the day with Georgetown's defeat of a Denver team and a "nine" from Golden. Occasionally, too, someone set off an explosion of "giant powder" to start the day with an echoing roar at dawn. Back in 1922, the *Courier* was pleased that "no one was disturbed by the explosion."[14]

The *Courier* showed, on another occasion, that times were changing, however. It issued an apology for "neglecting to publish the program and announcement for the fourth." Why? "The fault, however, is not all ours. When we—editor, reporter. compositor, and devil—are all one man it is more work to keep up and we are apt to overlook details." That Georgetown still had a newspaper proved noteworthy, even if a one-man show. Many of its contemporaries did not.

Another nuisance of the nineteenth century still wandered around. That was the burro, whose eating habits and determination were legendary. Georgetownites were warned that "no gate or fence is protection against their depredations. A few nights ago they did damage of the amount of several dollars by breaking their way through the door of a building." The article continued, "Now that people are getting their yards in shape and preparing gardens, owners of jacks should keep them off the streets." Cattle, too, caused a few moments of concern, particularly, as in 1924, when they were pastured in the cemetery: "no doubt someone thinks it a crime to let perfectly good grass go to waste when his cattle need nourishment."

Republicans repeatedly swept the town and county elections. Prohibition lurked in the background in almost every race. The paper got quite a chuckle out of a prohibitionist Democratic candidate in 1924 who "got an overload of tangle foot while electioneering."[15]

The editor tried hard, as in the 1928 presidential election. "For President Who But Hoover," the newspaper repeatedly and enthusiastically proclaimed. Proudly, the next March, with his inauguration the *Courier* expressed yet another hope. Herbert Hoover had been a professional mining engineer and an excellent one. "It seems probably and probable that he would lend a more sympathetic ear to the problems of mining men than to any other industry."[16]

A higher tariff on lead would be an immediate help to Colorado mining, the paper concluded. Thanks to cheaper labor, Mexican lead mining had seriously undermined the local industry, or so its supporters had claimed for years. Unfortunately, President Hoover, despite his mining background, did not have time to help the industry before a far greater calamity happened to the entire country.

Like a small-town newspaper at any time and place, the forte of the *Georgetown Courier* rested with its reporting of local news, those events that altered and illuminated readers' lives. That was what readers looked for on the front page, week after week.

It might be such an item as the weather being exceptional. "For the benefit of those who think that the Rocky Mountain section is the world in the winter please note that while the rest of the country has been suffering from cold we in Georgetown have not as yet had any zero weather." Or, the item could be concerning one of three remaining local veterans of the Civil War who had died.

When the playground committee of the PTA purchased a giant slide, emphasizing its being "extremely popular with children," that rated a story. So did their regular meetings during the school year, including such discussions as "serving milk and crackers every morning to children under weight is a splendid thing."

The Opera Club was a regular news item during its fall and winter meetings. Members at the "delightful meeting" of March 27, 1926, listened to a paper on the life of composer Richard Wagner, studied the opera *Tannhäuser*, and heard several instrumental and vocal selections. All told the "afternoon was most enjoyable and instructive." The Presbyterian church joined in sponsoring a series of lectures. Georgetown marched right in step with the rest of America with those ever popular clubs, some social, some instructional, and some both. The number of fraternal lodges had declined, but the Masons, especially, continued active.[17]

The paper also reported births, deaths, visitors, scout meetings, the opening of hunting season, church services, dances, parties, visitors, nearby community news, and that ever-popular topic, the weather. School news, including social activities, new teachers, programs, class projects, and high school sports appeared regularly during the academic year. The paper also published each month a list of the students with the highest grades in each class. Over what other students thought of these academic achievers, the *Courier* drew a veil of silence.

The public school continued to be a community focus, as it had been for decades. Pride in the school and its students never wavered. They were part of the glue that held the town together and gave it meaning.

The community's high point of the decade came with the construction of the town library. It did not prove that easy, though. Old-timer John Tomay, who had arrived back in 1873, died and left funds and property sufficient to erect a building. With buying and the donation of needed lots, all seemed ready to go in 1922. Several citizens, including the mayor, objected and went to court to try to break the will, for reasons the paper did not know or chose not to disclose. Two years dragged by in courts until the case finally reached the Colorado Supreme Court, which denied the attempt to break Tomay's will.

The library construction "progressed rapidly" in 1924, even when the discovery of "a jaw bone of an extinct animal" created quite a sensation. A minor flap occurred when some citizens protested the planned location of the toilets, but the architect refused to budge and would not change his plans. In the end, it all came together. November saw the library dedicated, with the laying of the corner stone included as part of a "fine program," featuring school children, several Masonic orders, firemen, and Idaho Spring's Blazing Arrow Band.

In January, the *Courier* published the rules adopted by the Library Association for the Tomay Memorial Library. Those included being open from 6:30–8:30 p.m. every day, except Sunday and also on Tuesday, Thursday, and Saturday afternoons. Each person could check out only one book at a time, to be kept just one week. Watch out for those two-cent fines for overdue books! Damaged books had to be paid for, and "all penalties would be rigidly enforced by the librarian, who would deliver no book to any person delinquent in non-payment of fines."[18]

Tax-paying Georgetown residents were given an advantage in using the library. They could get a library card by simply "filling out an application blank," while nonresidents were required to deposit $1 and pay five cents for each "volume taken from the library."[19]

Among those who could not get a card were the resident mountain sheep whose visits to town generally created a comment or two. One thing noticeable during the decade was the decline in local business advertising, to be replaced, usually, by ads about patent medicines and other items that could be purchased through the mail.

Nor did the paper overlook Christmas. It always featured a long report about the school program in the auditorium, "well filled with expectant children and parents." Elementary and high school proud parents particularly appreciated their children's efforts, but what the children thought about learning a piece to recite or sing received no mention. "Excellent music," the *Courier* crowed in 1922, from "home talent." Of course, Santa Claus always "jingled into the room and distributed good things from under the Christmas tree." Some years the celebration included a dance after the all important distribution of the gifts and candy.[20]

During the winter season, excited youngsters "took advantage of the fine coasting on Third Street." The four-block-long track "is a swift one."[21] Summer found the Huck Finns and Becky Thatchers of the 1920s socializing, fishing, hiking, playing, and generally having a good time, with thoughts of school far beyond.

For their parents, though, times were not good. Times had not been good when the decade opened, and plenty of indications appeared that they had not improved much in the roaring twenties. From the loss of population to the ending of railroad passenger traffic, everybody watched the deterioration of local conditions.

Nationally, signs pointed to the same thing. The farmer had not enjoyed the 1920s, and industry had overexpanded, resulting, by decade's end, in it facing large surpluses of goods. It had been too easy to borrow money for homes, cars, or stocks, and installment buying had become all the rage. Americans carried more personal debt than ever before, as purchasers cheerfully mortgaged their futures to obtain goods that would be consumed before they could be paid for. The saving and prudent spending virtues of earlier generations fell before the self-indulgence of flappers, bathtub gin, and the "good" times of the 1920s.

It did not take the stock market crash, with its infamous Black Thursday (October 24, 1929) on Wall Street, to tell the remaining Georgetown residents that times were going bad. At first, they had tried to make the best of the situation, like the "Hard Times" dance back in October 1927, at which "fun would be the watchword."[22]

Fun was not the watchword after October 1929. Georgetown had seen panics and depressions before—in 1873, 1899, and 1907—but nothing of this magnitude. As the year ended, the town and its people had slipped into a black economic, political, and social abyss. No way out brightened the horizon. Even optimists became pessimists amidst the gloom that settled over the country.

Hope had sprung eternal during the decade, but Georgetown's inhabitants, and many others, had forgotten what Ben Franklin warned his almanac readers 175 years earlier, "He that lives upon hope will die fasting." Dark days loomed at hand, Georgetown would never be the same again.

PHOTO ESSAY 7

# THE 1930s AND 1940s

*Georgetown High School students pose with "Indian clubs" during an exercise session at the old Georgetown School. Denver Public Library, Western History Collection x1283.*

*Georgetown men's basketball team pose outside the school c. 1930. Games were usually played at the Auditorium at the corner of Seventh and Rose Streets. Chappel family collection.*

*This photo of the Civilian Conservation Corps members at the Berthoud Pass Inn, is labeled "Mess For The CCC" and dated February 11, 1937. The men were probably part of the crew searching for two German skiers buried by a snowslide February 7. Bradley collection.*

*Young Edna Rydlund took this photo from the Silver Creek trail looking north toward the Alvarado Cemetery and Empire Junction in 1927. Alvarado Road is the main thoroughfare, with the railroad grade at the foot of Douglas Mountain on the left. In later years, as both US 40 and Interstate 70 came through this area, the bucolic nature of this photo would disappear forever. Edna Rydlund collection.*

*John Catren, Jesse Clair, and Dick Chappel pose at the entrance to the Baltimore Mine west of Silver Plume about 1927. Mining operations had certainly changed in scale, but there was still ore in the mountains and individuals willing to bring it to the surface. Chappel family collection.*

*Mount Evans CCC camp near Echo Lake. Bradley collection.*

*The Watrous Mill stood on the hillside southwest of Silver Plume, and is shown is this picture from about 1927. Dick Chappel, second from left, is showing the operation to "Mr. Briler" of the Denver Equipment Company and two Russian visitors. Even in hard times, many believed the industry could survive if they only had a new and better mill and some good investors. Chappel family collection.*

*While the original of this photo was labeled "Mount Evans CCC camp," it appears to be a Forest Service cabin with boy scouts and their leaders. Bradley collection.*

*Mr. and Mrs. Charles Johnson are perched on the U S Meridian Monument on the east side of Georgetown, mugging it up for the camera. The photo was taken c. 1930s. Denver Public Library, Western History Department f38166.*

*The Georgetown basketball team poses for a picture with their "trophy," the game ball, after placing second in a tournament in 1933. Chappel family photo.*

*The entrance to the L. E. Lemen scout camp west of Bakerville may have had a sign made by the scouts themselves. Scouting became a popular pastime for local boys and visiting troops from around the state. Chappel family collection.*

*After the end of passenger traffic on the Colorado and Southern Railroad in 1927, bus service became a lifeline for Georgetown and other mountain communities. Image courtesy of Historic Colorado (James Grafton Rogers collection, 10039779).*

*This postcard of the "Famous" Hotel De Paris is dated 1937. Plate lunches were selling for fifty cents, according to the sign. Bradley collection.*

# CHAPTER 8

# "TIMES ARE GETTIN' HARD, BOYS; MONEY'S GETTIN' SCARCE"

> Mamma taught us not to go outside eating something if we didn't have enough to share. If one of us would forget, and be eating an apple or an orange, you could bet that one of the neighborhood kids would say, "Dibs on the core" or "Dibs on the peel."
>
> *Rutherford family story*

The crash of 1929 did not immediately bring gloom and despair to the Clear Creek Valley. The Depression settled over the whole country at a varied pace. The impact within months, or at best a year, appeared in Georgetown and Silver Plume, and at their mines. A typical situation was that of the Mendota Mine, which did not operate in 1930, or the Griffith Mine, which listed "no production" the same year. The Centennial Mine, three years later, could only claim that it produced "a small amount of ore," with three miners. The Centennial Mill, which had been built in 1929, soon turned to producing chicken scratch.[1] The same year, the Commonwealth Tunnel had three men working it, who produced a "small amount of sacked ore." The three were "interested in a lease."

The market crash brought the rest of the country into a world already occupied by the residents of Georgetown. By 1930 only 300 people remained of the 3,500 who once inhabited the booming mining community. Unemployment, underemployment, sharp declines in property values, and investments slipping through holes in financial pockets had been commonplace for years.

Many of those who remained knew that their beautiful old homes, while probably in need of a coat of paint, still allowed families to live in a level of comfort and dignity that might not be possible if they sold, at a greatly reduced rate, and tried to reestablish their lives elsewhere.

The Board of County Commissioners urged people to pay their annual property tax but rarely put properties up for tax sale. While the sales would have generated revenue, it would have left many elderly widows homeless. For the next twenty years, Kneisel and Anderson grocery store wrote up bills for many of these older residents understanding that they might never receive payment.

Those who moved often stayed in contact with local friends, sending news of reunions and family events. The *Courier* noted that there were more Georgetown residents living in California than in the town itself.[2] Pioneer associations, picnics, and celebrations of nineteenth-century events filled the front page of the local paper along with a steady stream of obituaries.

Mining once had been Georgetown's present and future. Now times were changing. Georgetown faced a new crisis and with it changes unseen since the 1890s. Perhaps the letter to the Colorado commissioner of mines from Matt Sidney, March 2, 1930, best explained the plight of Clear Creek mining. After expecting work to start "next spring," throughout the previous decade, he had this to say: "The Sidney Tunnel Mining properties were owned by the defunct East Argentine Tunnel, Mining, Milling, Power & Transportation Co. [that covered all the bases!] and being the only surviving director it leaves me trustee. I have kept this property intact at my own expense and time without recompense to date. Work done during last year was in maintenance and re-timbering."

All told, it was not a very encouraging group of reports to show district production. Small operations, leasing, nonproducers, little production—all of it summarized perhaps by one leaser, who said he knew nothing about the mine's history or anything else, except for his small lease. Mining limped along while the years raced by.

Another problem that had to be faced was that smelters around the state were closing. The American Smelting & Refining Company responded to the changing times by shutting unneeded and unprofitable works. For Georgetown and Silver Plume miners this meant trucking their raw ore, and concentrated or milled ore, to Leadville for final reduction. That added both cost and time.

After fifty years, the Bank of Georgetown suddenly closed its doors on Wednesday, November 9, 1932. The bank's Board of Directors decided to take the action themselves, voting to put the operation into the hands of the State Bank Commission and to liquidate all assets.[3] The notice posted on the bank doors sent a shockwave through the community, although many had already removed their deposits over the previous year as banks had started to charge for services. In March 1932 the *Courier* noted: "President Hoover's request that the hoarded money be deposited in banks, should be coupled with a request that banks cease making a charge for taking care of money. This innovation on the part of many banks a year ago, caused the withdrawal of a million dollars from the Denver banks alone. Charging for taking care of money, with no guaranty that the owner will ever recover it, is an incentive for hoarding."[4]

Those who had money may have started hoarding prior to the bank closing its doors, but many Georgetown families simply did not have enough money to justify any kind of account. At the time of closure the bank was handling 156 accounts.[5] Local merchants and families struggled as their limited resources became entangled in a lengthy legal battle. The stockholders of the bank were personally liable for twice the amount of the stock they held. At the time of closure, bank stocks were owned by Marguerite Seifried McFarland ($11,000), Ruth Seifried Patterson ($10,000), J. B. Foley ($8,300), Howard Bushnell ($500), and Robert Christenson ($200), with a total value of $30,000. The Seifried sisters immediately recorded a mortgage on the family home on Rose Street and bank commissioner Grant McFerson delved into the liquidation of all bank-held assets.

Fear and uncertainty permeated the community. No one knew how long it would take to get money from the bank commissioner or what percentage of funds might be lost. The Bank of Georgetown's balance sheet showed $100,911.98 in assets and liabilities, but this included ownership or mortgage interests in properties that had lost value as the local real estate market plummeted.[6] Local families scrambled to "liquidate" their own assets in an effort to secure cash. The Randall family placed an ad in their own newspaper offering the extensive, treasured, Randall mineral collection for sale. At the same time, a call for calm was issued: "All people and all organizations will suffer the hardships that have come locally, in addition to the general hardships prevailing everywhere. But the spirit of faith and hope and cheerfulness will help everybody. We shall try to promote that spirit in our church services. And please, let everybody feel free to come to the church whether you have a million or a penny–either or neither you are welcome."[7]

Over the next year, depositors regained approximately one-third of their assets in three payments. The bank commissioner sold properties owned by the company, filed suit in local courts, and continued to work to reimburse bank patrons. When the commissioner had at least $5,000 in hand (roughly 10 percent of the amount needed to cover the deficit), a dividend would be issued to the account holders. By April 1934, 55 percent of funds had been secured. In February 1936 an additional "dividend" brought the balance up to 75 percent paid. Final court actions and sale of properties in 1938 brought the accounts close to full reimbursement.

While the town reeled with news of the bank closure in November 1932, the election of Franklin Roosevelt came as an additional shock to many of the stalwart Republicans in the small mining community. While times had become more difficult at the end of the 1920s, many Georgetown residents took solace in the fact that President Herbert Hoover was a mining engineer who recognized the importance of the local industry even if he couldn't single-handedly return it to prominence.

By the end of November 1932 Colorado received notice that the neediest individuals would be provided with food and clothing as long as funds lasted. A local relief committee with representatives from across the county was appointed to work with the American Red Cross to receive and distribute goods.[8] While locals would have preferred funding for labor projects, the commodities were appreciated nonetheless. In December the town's residents were advised to keep their coal bins locked when the theft of coal from the Presbyterian church was reported.[9]

Local support for President Roosevelt gained momentum in summer 1933 as the details of his recovery plans reached the local press. The expenditure of $3 billion for public works, reduction of the amount of gold in the dollar, and lifting of restrictions on the export of unretorted gold amalgam all held potential for Clear Creek county residents.[10] Enthusiasm grew with talk of an international pact regarding the production of silver and possible increase in the value of the white metal.

By September the traditionally Republican pages of the *Georgetown Courier* came on board with the headline "Good Citizens Will Support Roosevelt

Plan." As the elements of the National Recovery Act and related employment projects came forth, the paper urged citizens to support the programs and resist the temptation to doubt the effectiveness of the program. Support of the programs took on a patriotic air, with references to the war on unemployment and the duty to support the government in its efforts.

The state legislature faced similar difficulties and began to consider increases in taxes to help fund relief for the poorest families and to find ways to stimulate the state economy. A new tax on automobiles was passed to assist in paying for improvements to the highway system. The Rocky Mountain Motorist branch of the American Automobile Association had fought the tax for a number of years, but gradually accepted the idea once it became apparent that it was going to pass, noting that "while the auto tax was assessed, the Legislature, on the other hand, initiated a program of road building, which not only will save the motorists thousands of dollars in reduced cost of operating their cars but will improve conditions for everyone and will provide work for thousands."[11] While the state-sponsored projects would pale in comparison with the later federal programs, they started quickly and began to put people back to work.

In November 1933 the county commissioners and other local officials traveled to Colorado Springs to learn about the new "Civil Works" program funded by the government. The county allotment was sixty-four men, with future positions scheduled to come to the towns.[12] The following week the paper reported a crew of ten men working on the Green Lake Road and others at work on the cribbing of the creek in lower Georgetown—cold work for late November.

As the statewide network of roads began to improve, the mining industry began its own transformation, spurred by the hopes of price increases under the Roosevelt administration. Production had plummeted, and employment had all but dried up, but faith in the minerals within the mountains of Clear Creek County never wavered. The smallest glimmer of hope could spawn a new group of eastern investors, encouraged by the few remaining mining engineers and knowledgeable experts, anxious to be positioned to take advantage of the next gold rush if, and when, one might arise.

A large mining enterprise in the nineteenth century might have employed fifty or more men while a twentieth-century venture might hire five or six men. If the nineteenth-century population of 3,000 could be compared to the twentieth-century population of 300, the impact of the local mining industry on employment might seem comparable to the remaining residents. Four or five mining companies hiring five or six men might well be a sizeable industry within a small town.

At various times through the late 1920s and 1930s the Mineral Chief and Commonwealth Mines on Democrat Mountain, the Pulaski and Centennial Mines on Leavenworth Mountain, and the Capital Prize and East Griffith Mines on Alpine mountain hired multiple shifts of miners as they cleared old workings and continued to dig tunnels to prepare the property for the next coming of economic growth. The Mineral Chief, Commonwealth, Centennial and Capital Prize all built mills. The Mineral Chief constructed a new tramway and the Centennial Mine operation, owned by the optimistically named Western States Mining Company, tunneled south into Leavenworth Mountain and north underneath several blocks within Georgetown.

The Capital Prize and East Griffith Mines completed an assortment of tunnels and shafts aimed at connecting the workings of the valley floor with the Comet Mine workings on the ridgeline east of town. At the same time, they moved to connect with the workings of L. G. Davenport at the southern end of the Capital Central lodes north of Silver Dale.

The love-hate relationship with the railroad company flared up in 1927 with the discontinuation of passenger service to Clear Creek County. The newly formed Metal Miners' Association, with members from both Gilpin and Clear Creek counties, met to discuss the problem in 1927. The County Commissioners had previously filed a letter with the Colorado Public Utilities Commission in an effort to force the Colorado and Southern Railroad Company to continue passenger service.[13] The railroad company ceased passenger operations on an "experimental" basis for one year, starting in 1927; however, they agreed to continue freight service on a limited schedule. The company cited the lack of riders and the fact that both residents and tourists preferred use of busses or their own vehicles. The final decision to abandon passenger service came in May 1928, while freight services continued.[14]

The amount of ore shipped from the district continued to wane, yet still provided some revenue to the workers of the region. Individuals or family members often leased a single shaft in one of the larger mines. The "Anderson brothers and [Mr.] Jones" managed to fill an ore cart with workings from their lease on the Centennial Mine. After shipping and smelting costs were deducted, they netted $500. If their hours of work had been computed, the earnings would be slim; however, anything that could provide $500 in the 1930s was a success beyond belief.[15] While individual workers might be satisfied with their haul, others continued to interest investors, who were often even more excited with any notion of success. The owners of the Santiago properties in the Waldorf Basin near the original strikes in the East Argentine mining district shipped 108 tons of ore to the smelter in Leadville with a net return of $777, which was enough to encourage them to build a new sixty-ton mill![16]

Shipping of freight continued until the tracks were damaged by floods in summer 1933. Representatives of the Burlington Railroad Company were willing to meet with local mining representatives, but they were doubtful if repairs could be made unless the local companies agreed to ship more by rail and less by truck. Without pledges to increase the use of the rail lines, the company representatives indicated they would file for abandonment. The *Courier* stepped up to support the railroad companies: "We need the railroad because we cannot afford to ship our lower grade ore to the smelter by trucks; because we cannot afford to pay the extra charges on coal shipped in by trucks; because we are going to need a lot of new mining and milling equipment soon and we will want it shipped by the only safe and sure way—by rail."

While trucks and busses gave the locals more options, the editor finished his comments with a succinct note: "We MUST support the railroad if we are to expect it to remain in business. . . . If the railroad is abandoned it is the fault of shippers who prefer to ship by trucks."[17]

Miners pledged to ship by rail, and the company repaired the tracks, but the fragile partnership soon stumbled as production faltered and the Depression took an ever-increasing toll on the struggling rail industry. By the end of the decade the line would be closed and the tracks removed. In February 1939 the *Courier* noted that the high bridge had been sold to Charles Sutton of Empire for $400 and that the Colorado and Southern would soon commence removal of the tracks between Georgetown and Silver Plume.[18]

With the Depression deepening every year, mining, like other businesses, suffered. In 1933, at the depth of the economic woes, only $115,473 worth of gold and $11,345 in silver were mined in the county.

Much as Georgetown folk had done earlier in the 1890s, they looked to Washington, as did their neighbors, for help. That seemed to be their only and last hope.

Help, however, was on the way this time. Once more Uncle Sam rode to the rescue as America looked to President Franklin Roosevelt and his New Deal program for aid, which seemed to promise something for everyone, including mining.

Throughout 1932 into 1934, little mining news appeared in the *Courier*, but lots of articles on other things did, with the following types of headlines—"Silver Hopes High," "Silver Victory Looms," "We Want Silver Restored," and "Gold and Silver Seem Destined for Higher Prices."

A June 1934 article, "Return of Bimetallism Will Bring Prosperity," sounded like 1893 all over again. "If the world was back on a bimetallic basis there would be plenty of hard money for all purposes." This would help Georgetown, as well as some of its "ghost" neighbors. The author pointed out, "in every mining state in the West there are many so called 'ghost cities' that were once prosperous silver mining camps."[19]

As part of the sweeping New Deal reform program, gold miners received a major boost. In 1934, as a segment of his monetary policy, President Roosevelt took the country off the gold standard. When Congress passed the Gold Reserve Act, several things happened. Gold exports were forbidden, and gold coin, gold bullion, and gold certificates were taken out of circulation. What aided mining, the price of gold, was established at $35 per ounce, up from $20.67.

One thing almost immediately became apparent—placer mining returned to popularity especially around the early sites of placer discoveries in Clear Creek and Gilpin Counties. A sixth of an ounce of gold mined per day beat welfare payments or the low-paying jobs available for the numerous unemployed.

A Fifty-Niner would have been shocked to find out that the Public Works Administration sponsored classes in Denver to teach men and women the art of panning and placer mining. In and around Georgetown, some tried their luck with varying success. Nevertheless, it beat trying to live off the "dole," as they called it.

Western silver-producing states received a boost as well, thanks to Key Pittman once again. He helped to author, along with Montana's senator Burton Wheeler, the Silver Purchase Act (1934), which obliged the treasury to buy the entire output of American silver mines at a price above the world market price until the government stocks of silver reached one-third the value of the gold reserves. Silver miners received a generous subsidy with this act, which continued into the early 1940s.

The result of all these measures soon came into sight locally. Clear Creek mining clearly benefited, more so gold than silver. For example, in 1941, over $1,100,000 in gold was mined and $92,000 in silver. Lower Clear Creek received the biggest boost, but mining prospered as it had not in years.

The Santiago Mine, high above and southwest of Georgetown, was a typical operation for these years. Located at around 12,000 feet, its eight miners brought out fifty tons daily to be taken to the mill for concentration. Three men worked at the mill and two on the surface. The Bureau of Mines' inspector was satisfied with the mine's safety and sanitary conditions and those in the mill as well. He rated the "living conditions" good and all the buildings in "good" condition. Ore was trucked to Georgetown down a "fair auto road" and shipped via the railroad to Leadville.

Times changed quickly in mining during these days. The next year's report (1938) found ten men working the dumps, four in the mill, and none underground. The following year an improvement was made: the telephone arrived. Five miners were back digging, and no one worked the dumps.

More typical for these days were reports that appeared in the *Georgetown Courier* such as "A fine body of ore was opened on the Clara B Lode in the Commonwealth Tunnel" (July 1937), and it "is rumored that a ten year lease" has been signed on the Doric Tunnel (June 1938). In September 1938, Georgetown readers found out that a plan was afoot to "unwater the Centennial mine preparatory to an examination of the mine." They could also read that men were working the Anglo Saxon Mine and on Sherman Mountain.

"Mining Briefs," meanwhile, appeared less frequently in the newspaper, sometimes once a month, sometimes with a longer lapse. Mining was no longer what it had been in Clear Creek County. Nonetheless, the old hopes still existed.

That old-timer, the Griffith Mine, had fifteen men working in 1937, but "not producing," and one miner the next year with this comment again, "not producing at present." However the long-defunct Sidney Tunnel actually produced "a small amount of mill ore" with four miners working the mine in 1938. Sometimes hope springs more than eternal!

Travel by automobile and shipping by trucks got a boost from the economic stimulus expenditures of the Hoover administration and Roosevelt's New Deal as well as the state dollars generated from the first car license fees. The major overhaul of roads after the transition to automobile travel in the early 1900s had spurred further use of the roads resulting in the need for further improvements. Widening and paving as well as realignments aiming to reduce steep grades became commonplace.

The first state highways had been no more than a numbering system for roads that had received state funds to match county expenditures. In 1928 the state assumed total control of the "Federal Aid" roads within the state. County commissioners agreed to the change, but retained authority over the rest of the system, continuing the policy of supervising maintenance within the county boundaries and billing the state for one half of the cost.[20] In March 1928 the Good Roads Bureau of the Rocky Mountain Motorists association reported that 963.3 miles of Federal Aid roads had been completed in Colorado.

During the final days of the Hoover administration, in 1932, legislation passed authorizing hundreds of millions of dollars in road projects hoping to put people back to work, including "$120,000,000 for emergency construction on the federal aid highway system; $5,000,000 for national forest highways; $5,000,000 for improvements of national forests; $3,000,000 for improvements of national park roads, and $2,000,000 for roads through public lands."[21] At the same time, an additional $3 million was made available for roads as loans to the states through the Reconstruction Finance Corporation.

By the 1930s the cost of maintaining roads, even with loans or the 50 percent reimbursement by the state, overwhelmed Colorado's counties, especially a small county such as Clear Creek. The system of state and federal highways as well as the role of the United States Bureau of Public Roads, greatly expanded during the New Deal era, representing roads that would become the total responsibility of state and federal governments, a substantial restructuring. By 1933, Colorado had 9,136 miles in the state highway system with 4,203 miles surfaced with gravel and 467 surfaced with oil or concrete.[22]

Pressure mounted to keep the major passes open in the winter months, something unheard of during the horse and buggy days. In 1930 the government announced plans to keep Berthoud Pass open through the winter months. The Bureau of Public Roads released figures four years later showing that 53,000 cars had traveled across the pass in the winter months. Without snow removal, it was estimated that the pass would be closed eight months out of the year; however, revenue from the state gas tax, estimated at $25,000, made it possible to keep the road open.[23] In subsequent years, the ability to keep these roads open would facilitate the development of Colorado's ski industry.

Not surprisingly, the completion of federal roads increased the amount of interstate automobile travel, pulling more individuals away from the struggling railroads. While mining companies, Clear Creek County, and the communities within the county continued to stress the importance of shipping by rail, state and local forces shifted their interests to other priorities.

Tourists who once traveled to the mountains by rail, began to think about cars and the freedom to chose their own routes. The rapidly growing members of automobile clubs started to explore the mountains, traveling roads designed for horses and wagons.

In 1929 Colorado became the first state in the nation to "adequately" mark its historic and scenic sites with the assistance of the Motor Club.[24] Once drivers began to head west, pressure mounted for newer roads, including more direct routes where possible. Plans for building a road west from Golden through Clear Creek Canyon surfaced in 1930. The initial route assumed discontinuation of all rail service and met with loud complaints from the mountain communities. Subsequent plans were drawn to accommodate both rail and auto travel, and discussion turned to determining which road would be best for travel west of Denver: Floyd Hill or Clear Creek Canyon.

The Bureau of Public Roads wanted the road to wind through the canyon, while state highway engineer Charles Vail preferred improvements to the existing route over Floyd Hill. In 1937 Vail returned from Washington with approval for expenditure of over $1.5 million for construction of a four-lane highway over the Floyd Hill route and construction of a twenty-four-foot highway from Black Hawk to the forks of Clear Creek (the junction of today's US 6 and State Highway 119) then sharply west to the proposed junction with the Floyd Hill Road. While not the route envisioned by the early planners of the Bureau of Public Roads, it would at least start the canyon route and leave options open for later years.

In addition, the new route could replace the Virginia Canyon Road north of Idaho Springs as part of the proposed Peak to Peak highway, and also provide better access to Central City and Black Hawk. Patrons of the Central

City Opera (opened in 1932), including some of the most influential families in the state, had been traveling over the Virginia Canyon or Guy Gulch Roads on their way to the opera. They may well have been part of lobbying efforts to build the new stretch of Clear Creek Canyon Road.

Clear Creek County benefitted from the numerous road and forest projects. Enhancements to existing roads such as Floyd Hill, Berthoud and Loveland passes, as well as improvements to roads and bridges in between, changed a web of narrow, winding dirt roads into multilane highways suited for interstate travel. Support for the Hoover and Roosevelt road programs remained strong until the election of 1936, when the *Georgetown Courier* resumed its support of Republican candidates. An ad paid by the Republican Party pointed out that the New Deal projects had resulted in tremendous levels of debt: "THE POOR MAN'S BURDEN UNDER THE NEW DEAL! The New Deal has squandered MILLIONS! Debt has been piled up for you at the rate of $24,000,000 EVERY DAY. YOUR national debt is now 34 billions of dollars!"[25]

Concerns about debt weighed heavily on many, while others focused on the need to provide work to local families slipping into poverty. Jobs available through the Works Progress Administration (WPA) and Civilian Conservation Corps (CCC) made it possible for some individuals to support their families, though many had to work far from their homes as jobs opened across the state. Georgetown's William Rutherford accepted a supervisor's job with a forestry camp in Gunnison County, only to be laid off and transferred to a forestry camp near Pueblo. While he would return for holidays whenever possible, Rutherford also kept in touch with his young family by writing a series of stories, the *Whoopenhollar Kids*, with his children as leading characters. While the distance was hard on the family, the need for employment and the ability to provide for a wife and five children superseded all other considerations.

County officials continually bemoaned their inability to have workers placed locally; however, jobs anywhere were still valuable. Projects within the county included roadwork, construction of campgrounds, interviews with elderly residents, and inventories of local records. In 1937 the Forest Service applied to the WPA for funds to treat 25,000 acres of land in the Bear Creek Basin for damage from the "Clack Hills Beetle."[26]

In September 1937 the Georgetown school year began with thirty-one students in the high school program and fifty-seven in the elementary grades.

Efforts were made to incorporate programs to make graduates more employable, with bookkeeping added for one year and shorthand the next.[27] The large rooms and high ceilings of the 1874 brick schoolhouse made the building hard to heat, and parents began to grumble.

The following summer, members of Georgetown School District 3 met with representatives of the WPA to see if they could qualify for a school construction program that was part of the New Deal. The WPA would provide 40 percent of the funds if the community would agree to fund the remaining 60 percent, requiring passage of a bond in the middle of the Depression. The *Courier* set aside its conservative fiscal ways to urge the acceptance of the program: "The Georgetown school was built in 1874 and has served as such for 64 years. It is old. Money cannot repair it and make a modern building of it that the people could be proud of—for who is really proud of our public school building?" Many seemed convinced that the 1874 school was not long for the world: "Can we afford to reject a new school building costing $60,000 for less than $35,000, when the building we now have will only last a few more years, and then we will have to bear the entire cost of construction."[28] Not all were convinced. The next week longtime resident Frank Maxwell balked at the cost and advised against of adding debt during hard times:

> People find it difficult to pay their taxes now and if you add $1000 or $1500 a year for the life of the school bonds it will necessarily result in more people becoming delinquent, because of inability to pay. The R R will be removed long before the bonds are paid and their proportion of the tax will be added to yours.[29]

The editor responded with the "Other Side of the Question":

> The spending of approximately $50,000 for a new school building in Georgetown, while it will not all be spent in Georgetown, will have a tendency to uplift the morale and increase the purchasing power of every material man and laborer who participates in the work. The intent of the Government's huge spending and lending program is to do just that: namely help the morale and increase the purchasing power of the people and it has already had a marked effect on the outlook, as evidenced by increased business activity and a remarkable turn in the stock market; why not become a part of the program for the betterment of local and national prosperity?[30]

Hard times aside, Georgetown agreed to fund their 60 percent of construction costs and joined Idaho Springs and Empire in the construction of new school buildings.

After a review of several building sites, the town opted for construction on the site of the old Barton House Hotel on Taos Street. While many preferred a location further north, where there would be more room for playgrounds, it was noted that the cost of extending water lines 200 feet and sewer lines 600 feet would be excessive.[31] Project architect Lester L. Jones recommended the Barton Hill site, and the School Board agreed with his proposal.

In October 1938 a telegram came from senators Edwin Johnson and Alva Adams: "Public Works Administration has approved Docket 1290 for grant $21,681 for School District No 3 Georgetown. Preliminaries are being rushed as rapidly as possible and construction will soon be under way."[32] Final plans were approved in December 1938, and David A. Olson of Denver was selected as the contractor.[33] Plans called for "three grade rooms, a high school assembly room and two recitation rooms and the superintendent's office and an auditorium and gymnasium."[34] It was hoped that a stage could be built within the auditorium. The contractor was encouraged to use local labor through the Public Works Office in Idaho Springs and to move quickly to assure that the building could be ready for the start of the school year in September 1939.

Despite a cold winter and six-week delay in delivery of brick due to a labor strike, the building opened for public inspection at the end of July. The new building, "up to date in every respect," drew praise from all, but the need for new equipment was noted, and a series of dances scheduled to raise money for better furnishings.[35]

The school opened its doors for the first classes on September 6, 1939, with thirty-four students in the high school and forty-nine in grade school. Adult classes were offered in shorthand and typing. In November a nursery school under the direction of the PTA was started under the direction of Mrs. Henry Anderson with the assistance of Mrs. William Hollingsworth. The nursery school's goals focused on teaching children to play with each other and included "sand box, clay work, coloring, cutting, weaving, songs and stories."[36]

As money gradually started to return to the nation's struggling economy, skiers and ski clubs began traveling to the upper reaches of Colorado's mountains. Interest in skiing, especially ski jumping, had surfaced in the 1920s. Ski meets increased in numbers, and the general level of interest spiked. Most of the Colorado colleges and universities soon had ski clubs that started to sponsor competitions. The international ski industry took major steps in the 1930s as ski manufacturers such as Austrian Marius Ericksen (father of Olympian Stein) and Coloradan Thor C. Groswold started to manufacture skis designed for downhill (ski-running) use, instead of the old-fashioned Nordic skis designed for travel across deep snow and flat terrain or the large, relatively flat skis designed for ski jumping.

The work on both Berthoud and Loveland passes, as well as the decision to keep the roads open during the winter months, added a great deal of vertical footage to the available ski runs. Skiers could drive to the top of the pass, ski down, and catch a ride back to the top. Most of the land was under the supervision of the US Forest Service, which willingly helped locate and site new runs. In January 1937 the Forest Service approved an application from the Denver Winter Sports Club to install a rope tow on Berthoud Pass, noting that "only a minimal fee will be charged to the public."[37] Later that month the Hotel De Paris hosted a "jolly bunch of skiers" over the weekend.[38]

The basin at the foot of Loveland Pass began to attract skiers at about the same time when a temporary rope tow was operated by J. C. Blickensderfer. Jonas and Clifford Bennett, miners living in Silver Plume, began clearing trails and running a lift designed to pull riders up the high slopes. The Bennetts told friends that if they could figure out how to get ore down the mountain, they certainly could get skiers up it.[39] In later years, a second family of Bennetts (Allen W. Bennett), unrelated to these two Silver Plume gentlemen, would be major players in the development of the Loveland ski area.

The Civilian Conservation Corps built a log structure at the foot of the pass which served as a warming hut for a number of years, though later burned down. The new runs and fledgling ski area would later be known as Loveland Basin.

In 1939 the Zipfelberger Ski Club of Denver, in coordination with forest ranger Wilfred S. Davis (who had recently returned from Switzerland), announced a race, on April 16, starting at the top of Loveland pass, then skiing down the east side along a new, recently discovered trail with 2,258 feet of drop along a two-and-one-half-mile stretch. The Zips (as the club was locally known) also applied for a permit from the USFS for construction of a cabin at the foot of the new trail.

The search for new trails and ways to access them must have resembled a new rush to the Rockies with major implications for the future of the state; however, the clouds of war would soon put a halt to the expansion of skiing. Many of the young skiers would train for battle, leaving behind the enjoyment of intercollegiate competition.

At the end of the 1930s, however, the local press discounted the possibility of war, and complained mightily about the huge debt piling up as part of the New Deal (the brief support of FDR having lapsed after one or two years). In an article entitled "Keep Out of War," the editor fussed at the president: "In the last ten days the country, quite justifiably it seems to us, has become alarmed over recent utterances of President Roosevelt which have had the effect of creating a war psychology in this nation. As indicating the frame of the Presidential mind, Mr. Roosevelt's words have been as frightening as they have been amazing."[40] The call for isolationism, similar to the leadup to World War I, would soon fall on deaf ears.

PHOTO ESSAY 8

# THE OFFICE OF WAR INFORMATION DOCUMENTS THE TOWN

Between 1935 and 1944, the Farm Security Administration–Office of War Information hired photographers to document towns and cities across the country. Director Roy Stryker hired a handful of photographers to travel the country and document the lifestyles and buildings of the American people. Over 170,000 photos were taken during the nine years the program was in existence. The negatives have been carefully preserved and indexed by the staff of the Library of Congress.

Photographers Marion Post Wolcott, John Vachon, and Arthur Rothstein traveled to Georgetown and Silver Plume between 1939 and 1942, capturing images with a clarity unseen since the days of William Henry Jackson's trips to the area in the nineteenth century. Few individuals show up in these photos. It is unknown whether this was a preference of the photographers, or an indication that there simply weren't many people in town.

Rothstein came to town in October 1939, Post in September 1941, and Vachon in May 1942. Rothstein seemed to be fascinated with the Hotel De Paris. A few interior shots of the hotel are also available from this time period.

*A John Vachon photograph taken in May 1942 shows the back of the county courthouse and the view to the southeast up Guanella Pass. The old town jail (small stone building) is shown in at the end of Fifth street in back of the courthouse, in its original location. Library of Congress, Prints & Photographs Division, FSA-OWI collection LC-USF34-065671-D DLC.*

*Marion Post Wolcott took this photo in September 1941. Only a few cars can be seen on Sixth Street. Library of Congress Prints & Photographs Division, FSA-OWI Collection LC-USF34-059132-D DLC.*

*The original county courthouse with its brick addition, is shown in this John Vachon photograph taken May 1942 as part of the Depression Era Series. The old town jail peeks out from behind. Library of Congress Prints & Photographs Division, FSA-OWI Collection LC USF34-065672-D DLC.*

*Photographer Arthur Rothstein must have wondered how the residents of Georgetown found jobs in what could only be called a sleepy economy in October 1939. A total of three cars and four people can be seen on Sixth Street. Library of Congress Prints & Photographs Division, FSA-OWI collection LC-USF34-028799-D DLC.*

*Silver Plume in September 1941 looked like a ghost town in this photo by Marion Post Wolcott. The mine dumps from the Diamond tunnel are in the background. Library of Congress, Prints & Photographs Division, FSA-OWI Collection LC-USF34-059056-D DLC.*

*The Cushman Block at the corner of Taos and Sixth Streets, sports a "Welcome" sign over the door of a largely empty building. John Vachon took the photo in May 1942. Library of Congress, Prints & Photographs division. FSA-OWI collection. LC-USF34-065674-D.*

*The buildings along Taos Street, later known as Taos Square, stand empty in this Arthur Rothstein photo taken October 1939. Library of Congress, Prints & Photographs division. FSA-OWI collection. LC-USF34-28843-D.*

*The beautiful detailing of the Ben Draper home on Taos Street caught the eye of Arthur Rothstein in October 1939. One of the quoins on the left side of the building has "Ben Draper" marked instead of a house number. Library of Congress, Prints & Photographs division. FSA-OWI collection. LC-USF34-28833-D.*

*The front door of the Star Hook and Ladder building stands open to show one of the towns hose carts. It would be another fourteen years before the town would stop using the carts and purchase a truck. Photo taken October 1939 by Arthur Rothstein. Library of Congress, Prints & Photographs division. FSA-OWI collection. LC-USF34-28838-D.*

*The commercial district along Sixth Street is languishing in this Arthur Rothstein photo taken in 1939. A lone sign saying "EAT" is the only sign of life. Library of Congress, Prints & Photographs division. FSA-OWI collection. LC-USF34-28835-D.*

*Smoke from the rooms in the back of the Hotel De Paris show a sign of some life in this Depression-era photo taken October 1939. Photo by Arthur Rothstein. Library of Congress, Prints & Photographs division. FSA-OWI collection. LC-USF34-28832-D.*

*Another view of the Hotel De Paris on Sixth Street shows the stucco work on the west side of the building starting to fall. Depression series—Hotel De Paris in Georgetown, October 1939. Photo by Arthur Rothstein. Library of Congress, Prints & Photographs division. FSA-OWI collection. LC-USF34-28798-D.*

FACING PAGE. *And one more photo by Rothstein, who may have wanted to make sure that the scales of justice on the roof of the Hotel De Paris got shown clearly. Hotel De Paris, Georgetown in October 1939. Photo by Arthur Rothstein. Library of Congress, Prints & Photographs division. LC-USF34-28796-D.*

*Photographer Arthur Rothstein appears to be fascinated with the beautiful old building. The photo gives an excellent study of the various patterns in wirework above the porch and balcony added by Sarah Burkholder. The photo dates from October 1939. Library of Congress, Prints & Photographs division. LC-USF34-28797-D.*

*The Capital Prize Mill and tailings, running from Biddle Street, across Main Street to Griffith, remained as a symbol of mining in the past, with few if any people employed by the mine at this time. Photo by Marion Post Wolcott, September 1941. Library of Congress, Prints & Photographs division. LC-USF34-059131-D.*

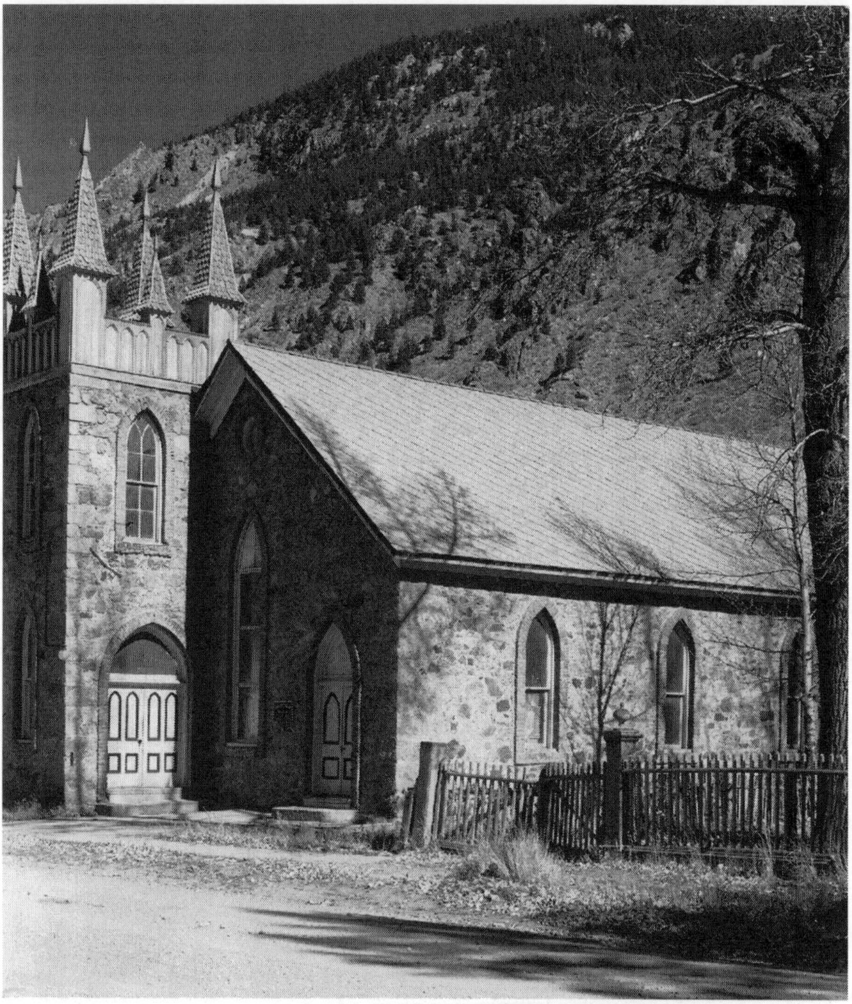

*First United Presbyterian Church of Georgetown, standing proudly next to the South Branch of South Clear Creek. The congregation continued to meet throughout the tough times, though sometimes parishioners brought coal for a donation on Sunday mornings. Photo by Arthur Rothstein, October 1939. Library of Congress, Prints & Photographs division. LC-USF34-028834-D.*

## CHAPTER 9

# GEORGETOWN GOES TO WAR

In September 1939, the world went to war, while America, for the moment, stayed out of the conflict. Within a year, though, the United States was placing itself on a war-preparedness situation. For mines that produced needed war materials, this ended the hard times of the 1930s.

War caught many by surprise. The slight uptick in the prospects for the mineral belt in Colorado's mountains had tempered the fear of war. Most felt that the United States would not join the war in Europe, hoping to avoid the casualties of World War I, the "war to end all wars."

The home front remained secure. Georgetown's families felt the Depression beginning to disappear into the past. At the start of December 1941, the excitement of the Christmas season held the attention of Georgetown's youngsters. Finally that exciting eve of Santa's appearance and day neared. Not only that, a school holiday loomed just around the corner. For the excited children, all this could not have come to pass too fast. Many already had their wish list in mind, and some had already talked to jolly St. Nick.

Georgetown residents opened their December 5, 1941, newspaper to check on local and general news. Among other items, they found that the regular meetings of the Students Literary Club and the Ladies Aid Group of the Presbyterian Church highlighted the front page of the *Georgetown Courier*.[1]

Sports fans read that, as expected, Joe DiMaggio was chosen the American League's most valuable player, after his fifty-six-game hitting streak the previous season. They could also check on the twenty-second and twenty-third installments of William MacLeod Raine's *Riders of Buck River*, as the story galloped on, keeping its readers on the edge of their seats.

For the ladies, the "buttoned up look in long waisted, snug fitting" dress presented the stylish hit of the season. The *Courier* offered Christmas cards for sale, and the weekly, editorial, "Feeling the Christmas Spirit," put the readers in the holiday mood.

All that aside, these times proved stressful and dangerous. While the hard times of the 1930s had eased a bit, the Depression still hung grimly on, despite all the legislative efforts of Franklin Roosevelt's New Deal. To make the season darker, the international situation looked equally grim. Nazi Germany seemed to be having its way in Europe and on the Russian front, and the Japanese were making war noises in the Far East and fighting in China.

Many *Courier* readers, no doubt, were not "feeling" so excited by the Christmas spirit, not with the war going on in Europe since September 1939. Militarily, the United States was working to get prepared. The army, for instance, showed "progress" in maneuvers in the Carolinas. However, in case war came, did they have enough time?

Army troops, meanwhile, had been sent to Dutch Guiana to cooperate with Dutch forces in protecting the bauxite mines there that furnished more than 60 percent of the American supply needed by the aluminum industry. As might be expected, navy enlistments had fallen off 15 percent following the torpedoing of naval ships in the North Atlantic by German submarines. Some parents no longer appeared willing to give their consent for their underage sons, younger than eighteen, to volunteer to join the navy.

That was the way it was in that early December 1941. Worries and fears reached even to the upper Clear Creek Valley. Before the *Courier*'s next issue came out President Roosevelt told the nation. "Yesterday—December 7, 1941—a date that will live in infamy—the United States of America was suddenly and deliberately attacked by naval and air forces of the Empire of Japan."

Georgetown residents must have obtained their news about the attacks on Pearl Harbor, the Philippines, and elsewhere from some place other than the *Courier.* The December 12 issue of the paper carried no war news on pages 1 and 2, which always featured the current local, national, and world news. One article told of secretary of state Cordell Hull talking with the Japanese envoys trying to find some path to peace. That seemed a little bit of late news.

The two longest articles dealt with rather interesting subjects, if the time were considered. One published the new ordinance prohibiting fireworks of all kinds and repealed all other ordinances relating to that topic. The other offered a well-illustrated, long story about "canine cops." A third featured ideas for Christmas gift boxes.

The only hint of trouble appeared in a short article about the Presbyterian Sunday service. The Rev. William Marshall preached on "God Understands," which proved a "source of comfort to his listeners." It would not be until the December 19 issue that the *Courier* finally, but only partially, caught up with the times: "Though the immediate attack came without warning the strong defense forces in Pearl Harbor went into quick action." The "Pearl Harbor news" stated that a large number of American planes had been destroyed along with one battleship and a destroyer that had been sunk.

Actually, on the sixteenth, the government finally provided some official information on the Pearl Harbor disaster, but the *Courier* did not carry it. Empire's correspondent, though, understood. The attack "thundered down upon us, we were stunned, shocked and petrified with horror." The war, they feared "had come upon us."[2]

Not until the next issue amid, "MERRY CHRISTMAS and Happy New Year" greetings did the *Courier* finally catch up with the times. It reported Secretary of Navy Frank Knox's news conference, which provided some gloomy statistics about what had happened at Pearl Harbor. With a never-say-die attitude, the article concluded with the encouraging statement that Japan had failed "in what it set out to do—knockout the United States in the Pacific before the war actually got under way." Like many others, the editor asked, "Was there any dereliction of duty prior to the attack?"[3]

Rumors too drifted around the town. They were partially right in that the attack planes "are believed to be operating from an aircraft carrier somewhere in the region." Not one aircraft carrier but six launched the planes, but it was wrong in reporting that one Japanese carrier had been sunk.

With the shock of Pearl Harbor barely wearing off, Georgetown folk purchased more war bonds and war-bond stamps than ever before. Nevertheless, people were "urged to buy more." They pushed not only bonds though. Among the various slogans was this one: "Victory Begins at Home Do Your Part."

They also sang World War II songs that aroused patriotism, if perhaps not likely to be long remembered once the war's end finally arrived. One such was "Remember Pearl Harbor."

> Let's REMEMBER PEARL HARBOR
> As we go to meet the foe.
> Let's REMEMBER PEARL HARBOR
> As we did the Alamo.

No doubt it encouraged locals and others in the dark days of early 1942 when seemingly little good news surfaced.

On the home front, Georgetown and its residents marched patriotically off to war. They, like Americans across the country, were angry, infuriated, "boiling" mad, and determined to do whatever they could to win the war. They read in the *Courier* to expect sabotage and perhaps the "chance of an air attack. No one knows when an actual bombing attack will be made somewhere."[4]

Unlike the recently closed decade of the 1930s, there was encouraging news amid the gloom: war-related jobs would be generated. Everybody would have to pitch in to help win the war. In the meantime all needed to remain particularly vigilant in preventing fires, so that they would not damage the war effort. To encourage everyone, though, the paper included photographs of the B-19 bomber, "Uncle Sam's aerial giant," and two P-40 fighters. America and Americans would win this war, no matter how long it might take.

People could not forget the day of December 7, however. It was not long before Georgetown residents entered into an argument over who was at fault for the Pearl Harbor disaster. The United States certainly would not have been caught so unprepared and unready unless an individual, or individuals, had failed. These citizens, like many others, came to blame the army and navy commanders, Lieutenant General Walter Short and Admiral Ernest Kimmel, for "dereliction of duty." Georgetown folk watched with interest as throughout the rest of the war and after, hearings and testimony continued to try to seek out the elusive truth.

One tidbit of good news started to come to the front. The December 7, 1941, attack on Pearl Harbor and America's entry into World War II, had caused an immediate impact on mining everywhere. The wartime needs for copper, zinc, molybdenum, and lead boosted Colorado production, as it did for coal. Those mines prospered as they had not for years; the Urad Mine, for instance, reopened in 1942, and molybdenum again became a major county producer. Jobs aplenty appeared, and miners also earned a draft deferment working in these mines.

Gold and silver, nonetheless, did not fall into essential wartime necessities. The result was the Gold Limitation Order, L-280, issued in October 1942. It closed nonessential mineral mines by not allowing operators access to replacements, or needed materials, and forbidding work other than maintenance. The purpose of L-280 was to free miners and equipment for the mines that were producing more critical war metals. Strictly gold and silver mines were ordered shut down, and mine owners suddenly became very interested in what had previously been the by-product of production. Of course, not all operators approved of such a policy; this to them proved just another "infamous act" passed by "that man," as the Roosevelt haters referred to him.

Actually, the order by itself did not suddenly put a stop to Clear Creek gold and silver mining. Even before the war's outbreak, miners began leaving to look for higher wages found in, for example, defense plants. By the end of 1941, precious metal mine operators also faced equipment shortages. In March 1942, before L-280 went into effect, transforming the mining scene, mines that produced gold and silver in excess of 30 percent of their total output lost their priority ratings for obtaining essential supplies and equipment. Order L-280 only completed a process already well underway in the county, state, and nation.

So Georgetown, Silver Plume, and the Clear Creek Mining Districts hastened into war while the county residents remained stunned with news from the front. Some mines benefited; others either closed or simply planned a maintenance program. Obviously molybdenum mines operated, and mines that produced gold and silver as by-products mined $158,480 in gold and $140,576 in silver in 1943, for example. Some county residents would be assigned to the town of Climax to work the molybdenum mine at the summit of Fremont Pass between Lake and Summit Counties.

War news had increased in the weeks that followed Pearl Harbor and offered little to cheer about. The Japanese empire seemed on the march everywhere. Georgetown readers followed the Philippine campaign most closely. There American troops struggled in combat. Nor had the war tide turned against Germany in Europe.

After being "heroically" defended, Bataan fell, and at long last Corregidor surrendered. In both cases the trapped defenders were overwhelmed by veteran Japanese soldiers, who had air and naval support. Americans could only read about this with frustration and anger. Then came the stories of the "death march," and attitudes toward the Japanese hardened even more. The *Courier* published an article declaring the "Jap culture offensive."[5] Bigotry ruled the state and local press.

In those first months of the war, times altered so fast and individuals' lives changed almost overnight; a few written snapshots, a potpourri of life and times, helps make Georgetown more understandable.

The United States weather bureau announced that "weather maps" and "weather forecasts" would not be published "for the duration" of the war. With a bit of humor, the *Courier* predicted that now local weather prophets would rely on "reheumatiz" and those old folk wisdoms, such as "red in the night sailors delight, red in the morning, sailors take warning."

With a shortage of teachers because of the draft and higher-paying jobs outside of education, substitute teachers could get "special war emergency certificates." Georgetown experienced this when popular Spanish teacher, Irene Ricci, left in late March 1942 to become a translator and interpreter for the War Department's intelligence division. Three months later, her "surprised" friends found out that she had "secretly married" her air force husband back on January 24. Not missing a beat, they sent their "congratulations and sincere wishes for happiness."

Almost as a sidelight, a more peaceful note appeared. Fishing season opened Monday morning, May 25, with a "grand rush to the streams and the lakes." Some "fair catches are reported."

As soon as war broke out, Georgetown men went into military service. Some local items appeared such as the fact that Mrs. George Horstmann "bravely but sorrowfully is giving her three sons for the defense of our country."

There were two messages that parents and friends at home did not want to hear: missing in action, killed in action. Georgetown in January 1942 had two young men in service, a number that jumped to twenty-four two years later and finally to twenty-nine by summer 1945. Unlike many communities, Georgetown's local men seemed to have led a charmed life.

Initially, the paper published servicemembers' names, ranks, and branch, but this stopped in 1943, when the Federal Office of Censorship developed new regulations: "Do not tell names of ships [that] sailors serve, do not tell units soldiers serve overseas. It is deemed inadvisable to publish the military unit and geographical location of men in service. We ask editors not to publish these troop identifications, and we ask parents and relatives not to reveal them." While it might not be a problem in Georgetown, the government worried about spies picking up useful tidbits of information at other locations from unsuspecting Americans. Henceforth, those details would be omitted in the *Courier*.

Georgetown residents, meanwhile, were humming or singing some of the popular World War II songs. Considering how many local boys went into the air force, one of the hit tunes, the 1943 "Coming in on a Wing and a Prayer" must have been all the rage.

> Coming in on a wing and a prayer
> Though there's one motor gone, we still carry on.
> What a show. What a fight.
> Yes, we really hit our target tonight.
> We're coming in on a wing and a prayer.

The words, though, soon took on a new meaning for locals as they did for all communities.

In the early years of the war, only one of the local men was reported missing in action. That was air force Sergeant James Marsh (April 30, 1943). Finally, in June, Marsh was reported to be a German prisoner. There was no further news until November, when his mother finally received word from him. He "had been shot down while flying over France," suffering a "badly burned face" and a "shattered ankle." Marsh told her that his face had healed but not his ankle.

Like many American boys, Marsh had never been abroad. He had written home back in August 1942, from "somewhere in England. England has been giving me a lot of experience that will be worth telling about when I get back to the States." Continuing, Marsh expressed a feeling of many of his comrades, "I never dreamed a year ago . . . I would be in Jolly Old England."

One of the first things Uncle Sam asked Americans to do was to take part in the "Food for Victory" program by planting gardens. Not simply were gardens deemed appropriate for nutrition, they also offered a further advantage: "in the interest of patriotism as well since any produce grown helped the war effort." "Everybody" certainly should grow a garden for food and patriotism, but "outside work helps keep you healthy," plus it provided you with needed sunshine and exercise. How many successful gardens Georgetown produced at its elevation (8,640 feet) and short growing season might be questioned.

Reading about the devastating air raids, Georgetown, like other towns, planned air raid drills, and the *Courier* published articles such as the "protection" required during raids. Like the rest of Colorado, the town practiced a blackout for a simulated air raid on December 14, 1942. "Turn off all lights," "arrange" to stay home, "refrain" from using the phone, and "all traffic except certified vehicles" was to stop. The ringing of eight "slow bells" and the firing of two signal bombs would indicate when the air raid drill started. Bells, after thirty minutes, would announce the drill's end. At the same time air raid wardens would canvass the town for lights being on and other "no-no's."

Georgetown residents performed their duties well, the *Courier* complimenting them for the "true American spirit shown by everyone obeying the rules without whine or question." Now, if the war came to the valley, they would be ready. In May 1943, they practiced again, this time with only a twenty-minute afternoon drill. Once more, all went well.

Whether it was because C. H. Randall faced financial shortages or paper shortages, the *Courier* shrank. For whatever reason, such as one of those wartime things that just happened, the editor decreased the size of his paper from eight to four pages in early 1942. Then in September 1943, the *Courier* put on a "new dress" and went back to eight pages. In its sixty-seventh year, the now easier-to-read paper featured sharper print and more columns, but the front page still contained local news. The rest of the paper consisted of war and state news, "canned" articles, and advertisements.

Randall did plead with his readers. Please pay the $1.50 yearly subscription, and, if "in arrears kindly pay up." What concerned him were the "expenses mounting," "income dwindling," and continuing static readership numbers. It proved an increasingly more difficult "task to keep things going." His problems, as pressing as they appeared, faced small town editors throughout the country.

The *Courier* started out by carrying war news to keep its readers as up to date as possible. For instance, on June 12, 1942, it proudly reported that at the Battle of Midway (June 3–6) two, possibly three, Japanese carriers had been sunk, while American casualties remained light. The six months after Pearl Harbor and the Philippines surrendering, plus other Japanese victories, provided a breath of encouraging news. Georgetown cheered the victory, and it proved a major one, with four carriers and a heavy cruiser sunk along with losing hard-to-replace experienced Japanese carrier pilots. Japan suffered a stunning defeat. Americans would not know for a while what their real cost totaled. The American Navy lost one carrier, one destroyer, and 113 planes, as well as pilots and seamen. Nor could they appreciate that this amazing victory, so quickly after Pearl Harbor, proved a turning point in the Pacific theater. Japan would never again launch an offensive campaign.

When the *Courier* went to four pages, war news shrunk to almost nothing, only to reappear when the paper resumed its eight pages. With Denver newspapers and the radio available, local war coverage was not that critical, however, as it had been in the past.

Patriotic Georgetown folks did all they could to help win the war. Within a few days after Pearl Harbor, they contributed to a Red Cross money drive to raise $50 million. How they might have responded to another Red Cross appeal, the newspaper did not say. The Red Cross had encouraged people to learn "how to keep our bodies and minds as fit as possible." In so doing, "we are helping the country, even as a very small unit."

They did not stop at simply raising money. The local chapter helped in the war effort in other ways. Red Cross members knitted warm socks for "soldiers, sailors, and marines" and "long wool scarves for aviators." By the fall of 1942 they turned to making "kits and bags for our boys in service" and organized a committee to "call on citizens" for cash donations to "buy equipment for these kits and bags."

The Georgetown and Silver Plume branches were also busy in the winter of 1943–44, making a total of 15 convalescent robes, 11 sweaters, 5 mufflers, and 3 "helmets." Despite all that effort, they appealed for more women to sew.

Women became involved in another way. The week of November 22–28, 1942, was designated as "Women at War Week." Now with women out doing men's jobs in factories and elsewhere, once again, as they had done in the Civil War and World War I, they proved they could do them successfully. With Georgetown women not being near any defense work, the *Courier* encouraged them to fill out their war stamp books and purchase bonds. "It is money that is going to win the war. Do your part by filling your war stamp books."

People were also encouraged to do their part in many other ways to win that war. These ranged from donating blood to writing letters. One article encouraged readers how to do the latter. "From his happy smile you can bet his girl will never lose him by poor letters! She's made her letter style as beguiling and smartly correct as her best date dress." Writing letters rode on a two-way street. Mothers of boys in service "radiate happiness to the post office when they receive letters from their beloved sons."

Everyone was encouraged to donate blood. "As you read the casualty lists, try and think, you could have helped keep some names off the list." The *Courier* also listed what to eat and not to eat six hours prior to giving blood.

In January 1943, locals joined in another "Victory Book Campaign." The "boys in the armed forces" need more "good books to read." Like patriotic folks everywhere, they responded. "Give more books, give good books, Georgetown gave a nice lot of books last year. Let us do still better now." When the campaign closed in March, the *Courier* proudly announced that the county had a "fine result," with 576 volumes. The paper especially congratulated the schools, churches, Red Cross, and other organizations which made it all possible.

Both Silver Plume and Georgetown also went out of their way to treat soldiers well, when they visited the communities and mountains. Silver Plume, for example, in early September 1943, served luncheon at the town hall for "almost 80" soldiers from Denver's Fort Logan and Lowry Field.

War Bond drives came regularly to the upper Clear Creek Valley. Reminders appeared in the newspaper. "Our job is to save dollars. Buy war bonds every pay day."

"Remember Bataan Invest a dime out of every dollar in U.S. WAR BONDS." On May 1, 1942, the paper reminded readers that the Clear Creek County quota for May was $3,900. "This is a quota for you and me to buy War Savings

Bonds and Stamps to the limit of our ability. It is also a quota in tanks and guns, planes and ships, torpedoes and bombs."

That system of giving advice to residents proved to be worth repeating. Locals also read articles about what things war bonds purchased. For example, the July 17, 1942, paper discussed cruisers that War Bond money helped build. Other articles explained how that money purchased a variety of war goods, including tanks, planes, and a variety of guns.

During the war bond drives, the *Courier* never failed to encourage readers to meet the Clear Creek County quota, which was $13,350, in August 1942. This urging never let up. In August 1943, the *Courier* pleaded, "it will take the combined efforts of every person in Clear Creek County to make our quota and we MUST make it." To enhance participation two four-foot bombs were on display in the county, and anyone who bought a bond may "autograph a bomb." They would then be returned to the ordnance and eventually "dropped onto Tokio [*sic*]."

Randall, in one of his rare signed editorials (September 10, 1943), asked the "people of this community, are you really at war?" The time for another bond drive had arrived, and he wanted to push hard for an increase in bond purchasing.

> What does being at war mean to you? Less gasoline for your auto? Less sugar in your coffee? High prices for your food and clothes? Separation from your husband? Sure, it has meant all these things, some serious and some really trivial for a period when your actual safety is at stake.
>
> The gold stars which decorate thousands of our homes today are a challenge to the vast majority of us to do our best in the Third War Loan. NO good American wants to see more gold stars added to the firmament of glory. We educate our youth for life, not for death. Every extra War Bond you buy today will back the invasion [of Europe] so that the war will be shortened and so that there will be less blood, sweat and tears for you and your friends.[6]

Like people throughout the country, Georgetown folks had to tighten their belts in many ways. Gas rationing limited their driving so "car pooling" and bus travel were encouraged. This brought up a humorous story. In February 1944 the town's barber came down with an attack of sciatica and could not work. "Many of our citizens have exhausted their allowance of gasoline and cannot spare time to travel by bus, and our hair grows longer and longer."

Politics went on as usual with local, state, and national campaigns. As the *Courier* pointed out, it was the first time since the Civil War that a presidential campaign had been conducted during wartime. There existed no question where the *Courier* stood: foursquare Republican. In 1944, FDR ran for a fourth time much to the dismay of Clear Creek County Republicans, who put their hopes on New York's Thomas Dewey. The *Courier* (September 22) complained that New Dealers launched a "near smear hysteria" campaign against Dewey "for his adherence to the fundamental principles of freedom," a campaign that "is perhaps without parallel." Otherwise, Republican Randall was silent in his newspaper except for "canned" Republican articles.

Local voters gave Dewey a majority along with the state and local GOP ticket, which must have pleased the editor. Listening all night on their radios, the minority Democrats of Clear Creek County cheered as the returns came in nationally. When it was finally announced, at 3:00 AM, that President Roosevelt had won an unprecedented fourth term, they breathed a sigh of relief. Disgusted editor Randall, however, was disappointed with the local turnout and the fact that a number of potential voters had failed to register.

Without question, Randall was ardently anti–New Deal and anti-Roosevelt by this time. He had once suggested that his readers give all of the new programs a chance, but quickly changed his mind. He muted his criticism after the shock of Pearl Harbor, and with everyone pulling together to defeat Japan and Germany, it was the patriotic thing to do. By late 1943, though, his criticisms of the president appeared quite frequently. It is interesting that these criticisms popped up in the middle of the local news columns, leaving the reader a bit confused as to their relevance there. Did he write them himself, or were they copied from some other source?

Not stopping, Randall also published longer anti-Roosevelt diatribes from Republican papers throughout the country. As the war went on, the articles only increased. When state elections rolled around, the Republican banner flew high.

Ration cards had now become a fact of life. So too had shortages. The ration board opened in the jury room of the courthouse. Here people picked up their "ration books" with coupons for purchasing sugar, for example, if any were available. Shortage of rubber meant that drivers could have only five tires; any others "must be disposed of by delivery to the nearest Railway

Express agency." No weather maps or forecasts were supplied "for the duration" in order not to give "aid and comfort to the enemy."

As mentioned previously, there was some "grumbling" about closing gold and silver mines. This "ill-advised ruling" was "made by some officious bureaucratic office holder." They closed nevertheless.

It might be war—and everyone was supposed to maintain their health—but, in the winter of 1943 (January and February), Georgetown suffered through a flu epidemic. It finally ran its course, but not before the paper carried weekly reports about how the flu sufferers were doing.

The United States was at war, but Georgetown residents and their fellow Americans tried, as best they could, to keep life as normal as possible. Life did go on in the upper Clear Creek Valley, amid all the other tribulations.

City government functioned, schools opened and closed, the town hall and the Alpine Hose House were painted, church services continued, and folks had to go to Idaho Springs to see the movies. The Loveland ski course was in perfect shape, and "many of the expert ski enthusiasts enjoyed dashing down the ski course," in the winter of 1942–43. Gas rationing and potential wear and tear on those hard-to-replace tires stopped some skiers from taking to the slopes. The ski areas were open on a very limited basis due to gas rationing.

The PTA met, as did the Boy Scouts, who conducted waste paper drives. Meanwhile, Cub Scouts and Camp Fire Girl groups were organized, giving Georgetown kids a variety of groups to join. People were urged to save scrap metal, which would be collected and "used in the war effort." A telephone crew, meantime, "is busy salvaging" old cable over Argentine Pass. Salvage operations at old mines yielded a variety of metal that ended up in war production.

The Presbyterian churches of Georgetown and Idaho Springs reached a milestone. Their minister, the Rev. William Marshal, proudly announced that "due to the excellent growth and strengthening," they had voted to discontinue receiving "aid from the Board of National Missions." As of May 1, 1943, both churches would be self-supporting.

The high school graduated seven students in May 1943. In spite of the war, the junior-senior classes held their prom, which featured a dinner at the Hotel De Paris followed by a trip to Denver and dancing at Elitch Gardens. Their graduation address was given by former governor Ralph Carr, who had gained fame by opposing a Japanese relocation camp to be built at Amache in southeastern Colorado. Many Coloradans were opposed to the idea of having those West Coast "traitors," in their state. Most of the interned, though, were actually American citizens. It took courage, which Carr had, to stand against the wave of criticism and "super patriotism" emotionalism that came his way and probably cost him his United States Senate race in 1942.

According to the May 28 *Courier*, Carr presented a "masterly" speech that featured a discussion of problems confronting the state and nation and a "clear portrayal of the solution of war problems and a peace setup that will of necessity need to be met and solved." It "thrilled everyone present."

For the youngsters, wartime had some unusual side effects. Teachers and parents talked to them about their Halloween activities, "We are patriots, and there will be no Hallowe'en [*sic*] pranks nor hand out calls for the duration of the war." The admonition apparently worked, as the *Courier* congratulated "our future American citizens."

Those future citizens' favorite season was probably Christmas, which the war impacted as well. Every year people were encouraged to mail their Christmas packages early, because of "unprecedented wartime demands" and a record volume of mailings. Also this hint: "war needs the wires please don't call war-busy cities this Christmas [1943]." Packages to men and women in service overseas should be marked "Christmas Gift Parcel and mailed by October 15."

There were fewer toys in the stores for youngsters to gaze at, admire, and wish for. Military toys predominated, and even dolls enlisted and came out nurses, Women Accepted for Volunteer Emergency Service (WAVES), and Women's Army Corps (WACs). Boys had a large variety of military toys to hope for. But stockings carefully hung up proved a little leaner during the war years. Christmas lights stayed on for shorter time periods, with burned-out bulbs and colorful Christmas strings and ornaments being hard to replace.

As a sidelight to the war, the number of marriages reported in the *Courier* definitely increased. Sometimes the *Courier* carried a couple of wedding announcements per month of Georgetown young men and women, considerably more than the prewar average. Their new homes were far away, as one or another, or both, now worked in defense jobs or served in the military.

As 1944 raced along, news from the war fronts became more and more encouraging. With Germany and Japan both retreating from their earlier conquests and Italy leaving the war, only to be propped up by German forces, the tide of war had clearly turned. That did not mean the Axis powers gave up; they still stubbornly fought Allied forces on all fronts.

Georgetown's good fortune of not having a local boy killed in action, however, came to an end. In late April, Lieutenant August Horstmann, and a month later Second Lieutenant Joseph Ricci, were both killed in the European theater. The heartbreak of these tragedies was shown in the article about Ricci in the February 2, 1945, *Courier*. "After eight months of torturing suspense, during which Mr. and Mrs. A. J. Ricci believed their son to be a prisoner in Germany, they received word last Saturday confirming the fact that he was killed in action on May 29."

Meanwhile, Georgetown folk continued buying war bonds, and contributing to Red Cross drives. The Boy Scouts gathered paper, and locals saved anything that could be used in the war effort. Following a Christmas seal drive for tuberculosis patients, the *Courier* observed that it had been a successful campaign even if the amount proved slightly less than the previous year. "In view of the decline in population since last year, it will represent a higher per capita contribution." The editor, therefore, found a silver lining in a less-than-promising future situation.

It was, and is, always hard for a newspaper to admit that the local population had declined, because growth had always been so important to western and Colorado communities. Like many other towns of its size, Georgetown could not provide the employment opportunities that larger cities offered, particularly in wartime. Thus the twentieth-century population trend continued downward for the mountain-enclosed community.

War-related problems appeared in a variety of ways. The telephone company delayed telephone installations, pointing out that after two years of war all the reserve telephone equipment "is used up." Military demands "use everything we can make including telephones, and equipment." The installations were used, for among other things, "secret electrical weapons."

In September 1944, popular Presbyterian minister William Marshall announced that he had been called to the church at Rawlings, Wyoming. He "would be greatly missed in our little church." An interesting question about all this comes from the fact that Georgetown had other churches, including Catholic, Episcopal and sometimes Baptist, but no mention of them ever appeared in the *Courier*, unless for a wedding or funeral! The Presbyterians, on the other hand, had a weekly column. As you might guess, the Randalls were Presbyterians.

The newspaper did continue to report on school activities, thanks particularly to high school student reporters. They showed enthusiasm, as this story of the January 25, 1945, basketball game against Nederland amply displayed: "The spectators sat breathless on the edge of their seats. The score 'rocked' back and forth in favor of one team and then the other. Finally the most exciting game of the year was over: Nederland winning 41 to 35."

The junior high students, meanwhile, spent time "learning how to dance." The *Courier* almost beamed with anticipation. "In a few years we are going to have some 'Arthur Murray's' performing at Georgetown High School."

Declining population, wartime shortages, and good and bad news from the war front did not prevent the community from celebrating Christmas 1944 in the usual way. Christmas carolers strolled about on Christmas eve and "were a joy to many citizens." The PTA earlier sponsored a community Christmas tree where all the children and their parents met Santa Claus. Then came the traditional school program, which "showed much preparation and was exceptionally fine." When it was over, each "child received the usual treat." The Sunday school class also "gave a splendid program" once again, with every youngster receiving a treat.

At the same time, Georgetown suffered through a "crime wave." After some one or several individuals tried to break into the pool room and with "other thieving going on for some time," editor Randall offered a suggestion. Parents should "see that their children are not running the streets at night. They may not be guilty of mischief but it leaves them open to suspicion."

Winter came late to Georgetown, as the war entered its fourth year, but, when it arrived it came in full force. Among other things, a Denver man was killed when a snowslide hit his car on Berthoud Pass. The *Courier* could only say in early April that "old man winter has been making up for it with a bang."

Snowbound Georgetown and the whole country suffered a shock when President Franklin Roosevelt died on April 12, 1945. Considering his previous anti-Roosevelt and anti–New Deal comments, Randall's eulogy included this amazing statement: "We can only say that one of the greatest men of the present time has been taken and his departure is mourned throughout the world. Future history will reveal the true greatness of this man." It raises the question of whether he, or someone else, wrote this unsigned hymn of praise.

Good news quickly followed. On May 8, VE Day, the European war ended. Then more good news reached town. James Marsh, liberated from prison camp, had returned to England, where he married his British fiancée, Peggy Doe. His Georgetown friends read that the wedding had been scheduled for June 1943. Then just a few days before that date his plane had been shot down, and he had spent the rest of the war in a German prison camp while his bride-to-be waited.

Marsh would return to Georgetown in September 1945 to visit his old friends. The *Courier* commented that "he looked fine" and had "regained most weight lost while suffering" in the POW camp. He told the paper that after his discharge he hoped soon to bring his bride home from England.

Summer 1945 saw the war still going on in the Pacific. War recovery had started in Europe; meanwhile signs of peace were sprouting. The newly organized United Nations conducted a nationwide clothing collection drive. Georgetown folk, along with Americans everywhere, contributed "used clothing, bedding, shoes, and other materials needed for overseas war relief." The American Cancer Society also launched a nationwide campaign, which they designated "Cancer Control Month."

The seventh war loan drive ended, and Clear Creek County again reached its quota, this time $98,000. That sparked the *Courier* to a moment of praise and an opinion, returning to its conservative Republican outlook: "This speaks mighty well of a county without a single war production industry and with mining, its only major industry, restricted by government bureaucrats."

The local troop of Boy Scouts, after all that work of collecting scrap paper, received a treat, earning a week at Camp Tahosa near Ward, "where they enjoyed horseback riding, hiking, boating, and trying to swim." The recently established Veterans of Foreign War post also organized a Ladies' Auxiliary.

Stories in the newspaper now told of servicemen and women returning to civilian life, the comings and goings of summer residents and visitors, and the hiring of teachers for the fall. Once again, all the men gained the title of professor, and the women received nothing more than just their names in the story.

The war against Japan ground on with particularly bloody battles at Iwo Jima and Okinawa, as the allied forces relentlessly moved closer to Japan. The Japanese fought fanatically and prepared for a desperate last stand at home.

The dreaded invasion of Japan did not have to take place after the dropping of atomic bombs on Hiroshima and Nagasaki on August 6 and 9.

The shocked world looked at the dead and wounded totals, and the damage done, in almost disbelief without realizing what the social, emotional, and medical aftermath would be. The destruction brought the Japanese government to realize resistance would be useless, and they surrendered. World War II officially ended on September 2, 1945, with the signing of the surrender documents.

Finally, the words of the popular song of the day, particularly the version featuring the English singer Vera Lynn, the "White Cliffs of Dover," came true.

There'll be bluebirds over
The white cliffs of Dover,
Tomorrow, just you wait and see.
There'll be love and laughter
And peace ever after.
Tomorrow, when the world is free.

A fascinating local issue covering the last weeks of the war showed that the *Georgetown Courier* made no mention of any of these events, except a "canned" piece or two on page 2 or 3. Not even a special notice of the war ending appeared on the front page.

Randall did observe on September 7 that servicemen "are starting to return to civilian life." He apparently had already reached the point of peacetime news. "The general exodus of summer visitors makes us realize that winter is not far off." Winter had not been far off back on December 7, 1941, either. Thus, Georgetown entered the new post–World War II world.

But what would come next? The town and the state of Colorado as a whole had endured decades of hard times. The crippling blow of the silver crash back in 1893, followed by a world war, the flu epidemic, a Depression of major magnitude, another world war, and food and gas rationing all left the residents of Georgetown uncertain as to whether anything would ever be the same. Could the mining industry recover? Even the largest mines, the steadiest producers now struggled to produce a product worth digging.

The Stevens Mine had bounced through a checkered career during the twentieth century. Leasers operated it when it was worked, but the mine was in bad shape. By 1945, as the war ended, the lower levels were filled with water, and the lessee was clearing the mine of ice and snow, planning apparently on retimbering, but "plans for further work were not definite." The once front-page mine had slipped a long way in the century, and it typified what had happened to local mining generally.

The Anglo Saxon Mine, two miles north of Georgetown, was one of those that had stopped operating. Once a silver mine with lead and zinc by-products, it did not have enough of either to continue. Nor did its location prove conducive to operating a small mine with low amounts of needed material. The May 1943 inspector's report simply stated, "There is a good auto road to within 1 ½ miles of mine, [the last] 1 ½ miles is a steep jack trail, supplies for mine are packed by jacks and horses." No production came from this property during the war, though a small maintenance crew kept the property in operable condition until it reopened after the war.

The Smuggler Mine just west of Silver Plume was one of the few local ones that continued operating during the conflict thanks to having zinc and lead. Once considered by-products, they had become the main reason for mining and outshone gold and silver. While it only had five men working there in 1941, and reported "not producing at present," the Smuggler was brought back to life by the war. By 1944, fifteen men worked underground, three on the surface, and three at their Silver Leaf Mill. With a good auto road to mine and mill, it had another advantage but still had to truck its ore to Leadville.

The Terrible Dunderberg Mine west of Silver Plume had also benefited from the wartime economy. Now operating as a lead, zinc, and silver mine just off US Highway 6, it had twenty-three miners working underground in 1945 and three on the surface. Its access, with an adit on the highway, proved beneficial for hauling and general access. It produced 600 tons of ore per month, as the war concluded.

As the guns fell silent, in August 1945, and the war ended, L-280 was rescinded. Peace proved no help to mining in either the state or Clear Creek County. Colorado gold and silver mining sat at its lowest ebb since 1873. It would take a while to find experienced miners, purchase needed equipment, and resume operations. Against this background, interest in exclusive gold and silver mining stayed low throughout the rest of the decade. The cost of mining had risen, as had miners' wages, and government regulations continued mounting, while the price of gold stayed at $35, and silver hovered low on the international market.

Mining never died in Clear Creek County, however, during the rest of the twentieth century. Small mining operations kept it alive in a sporadic manner throughout Clear Creek County. Two or three miners with a hopeful leaser, or owner, drifting along a small vein or reworking older areas was typical. It was not a lot of production by mining standards, only $129,360 in gold in 1946, along with $66,896 in silver. Three years later, as the decade closed, gold had sunk to $94,500, but silver staged a bit of a comeback with $76,020.

Hope never waned despite such totals. Somewhere up a rocky valley or on a high mountainside, there still nestled the mother lode just waiting to be found. Somewhere in a mine's depth, old-timers had missed a high-grade vein that was there now for the taking. Perhaps a new milling process would be discovered to work previously unprofitable ore. Perhaps some formerly unusable ore thrown out on a dump, like tungsten or molybdenum had been, would suddenly find a new use.

Maybe new equipment, or an invention, would drastically lower the mining costs. Maybe the price of gold and silver would rise, and precious metal mining would rise with them. The hopes were many, the realities harsh.

Colorado poet Thomas Hornsby Ferril caught that never-say-die spirit that had bedrocked mining for centuries in his poem, "Always Begin Where You Are."

Always begin right here where you are[7]
And work out from here:
If adrift, feel the feel of the oar in the oarlock first,
If saddling a horse let your right knee slug
The belly of the horse like an uppercut,
Then cinch his suck,
Then mount and ride away
To any dream deserving the sensible world.

So it had been since the days of 1849 and 1859, which scattered the ever-hopeful prospectors, miners, and investors throughout the West. Times might change, but peoples' basic instincts do not. Their frame of mind: that was what had kept mining alive since endless time-the lure, the search, the excitement, and the dream of wealth.

PHOTO ESSAY 9

# LIFE RETURNS TO NORMAL AFTER THE WAR

*The beautiful rock and timber workings inside the Burleigh Tunnel in Silver Plume made it a tourist destination. Chappel family collection.*

*Verona Chappel Zimmerman (on right) with friend, are standing with armfuls of columbines in front of a house on Main Street, Silver Plume c. 1940s. Chappel family collection.*

*By the 1940s Georgetown knew it would need the highway, instead of just mining, in order to survive. Two lanes of State Highway 91 are shown at the top of the photo, narrowing to a smaller two lane road as it approaches the center of town and the hairpin turns up the mountainside. Image used courtesy of History Colorado, 10026579.*

*This photo captures the county jail on Biddle Street a few weeks before it burned in 1946, with bars on the second-floor window. Denver Public Library, Western History collection x1228.*

*The Christian Science Hall (northeast corner of Seventh and Taos Streets) had closed its doors by the time this photo was taken c. 1940. The building was built for Morris's saloon, then purchased by the members of the local Christian Science Church. By the 1940s it had evolved into to a mattress-rebuilding establishment. Today the building has the sign "Georgetown Museum." Denver Public Library, Western History Collection x62975.*

*The brick jail and adjacent wooden building burned to the ground. Remnants of the building would later be converted into a private residence. Courtesy, History Colorado f33698.*

*Alpine Hose No. 2 looks reasonably solid in this George A. Grant photo taken October 1948. The property at the northeast corner of Fifth and Rose streets was donated to the town for use as a park, but would later be turned into a parking lot. National Park Service photo collection 10151, courtesy John Tomay Memorial Library.*

*Georgetown looks like a haunted village in this photo taken October 10, 1948, by George A. Grant. When compared to the 1939 photos of Arthur Rothstein, the town has picked up quite a few weeds, a couple of stop signs, and a big sign for a "bar" over the Fish Block building. National Park Service Collection 10156, courtesy John Tomay Memorial Library.*

*Even if mining had ground to a halt, the town still had a demand for food and drink. George A. Grant took this photo October 10, 1948. The new coat of white paint on the corner building, along with a sign for the Grub Stake Restaurant, are an indication that Ben Draper had purchased the property and was beginning to put resources into its restoration. National Park Service collection 10155, courtesy John Tomay Memorial Library.*

*Summer cottages became popular throughout the mountains in the 1930s and 1940s. This photo "The Cricket" was taken July 4, 1947, showing the "Cricket" cottage on Spring Street. Ronald J Neely collection.*

*The Ecklund mule train takes time to munch on weeds next door to the Hotel De Paris before heading up to the Argentine. The wooden buildings on the west side of Taos Street are empty and forlorn. Rutherford family collection.*

*This Georgetown photo from the 1940s shows a number of blocks in the commercial district. Several of the old wooden buildings have been torn down or burned. Rutherford family collection.*

*The Ecklund pack train heads up Fourth Street carrying supplies to the family mines on Crystal Creek in the Argentine Mining District. The long haul would require these animals to carry their goods up at least 2,000 feet in altitude. Rutherford family collection.*

*The remains of the Malm Mill are shown in this 1940s photo. The mill was built in 1910 with a great deal of enthusiasm on the part of the developers, echoed, of course, in the local papers. The processing system did not work out, and the building was sold to a wrecking company in 1919. Pieces of the concrete foundation remain in the vicinity of the town tennis courts. Rutherford family collection.*

*Young Buff Rutherford (age sixteen) working on the road up to Mount Evans in August 1945. As people began to drive their own cars to the mountains, Denver Mountain Parks and the US Forest Service decided to build a road to the top of Mt. Evans. Rutherford family collection.*

*The Crest House was built at the summit of the Mount Evans Road. At one time the route was part of the proposed Peak to Peak Highway. The road down the south face of the mountain was never developed. Denver Public Library, Western History Collection, z7794.*

*The early Shelter House is shown in this Forest Service notice. The Arapaho National Forest was a willing partner in the development of the Loveland ski area. Courtesy of Clear Creek Skiing Corporation.*

FACING PAGE. *The area's first Tucker Sno-Cat hauled skiers as well as performing a variety of other tasks that required the ability to traverse deep, snow-covered hills. This picture was probably taken in the late 1940s or early 1950s. Photo courtesy of Clear Creek Skiing Corporation.*

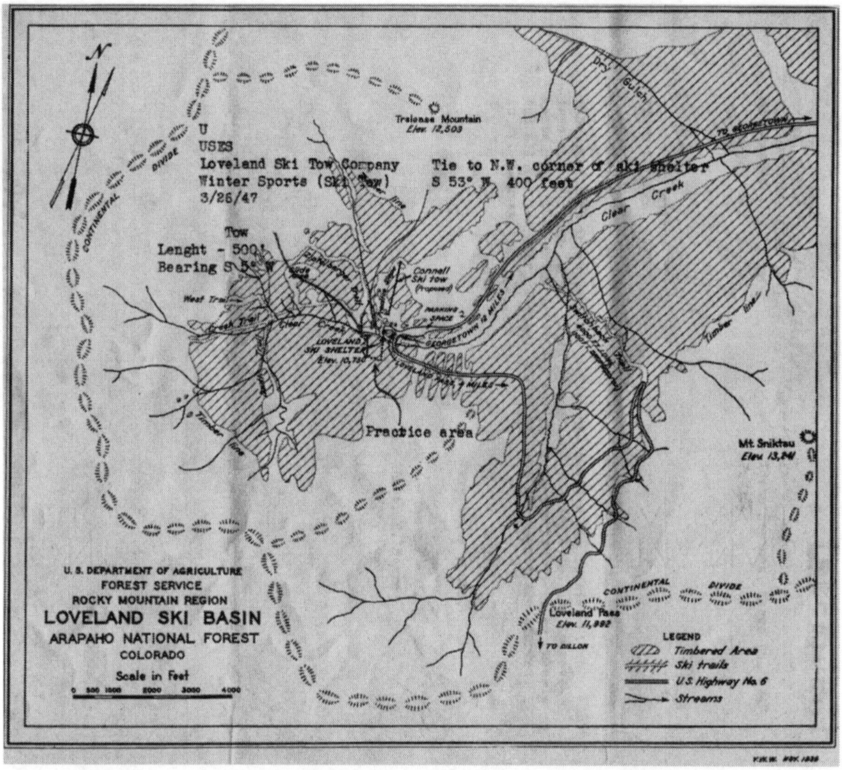


*Access to the early runs on the north side of the basin would be made more difficult with the construction of I-70. New runs were beginning to develop around the area in this 1947 map.*

*The early rope tow up the side north side of the valley is well remembered by many longtime Loveland skiers—all of whom tell tales of sore backs and greasy gloves. The early shelter building can be seen in the background. Photo courtesy of Clear Creek Skiing Corporation.*

## CHAPTER 10

# INTO A NEW ERA

World War II ended amid pilotless rockets and atomic bombs heralding a new era in military warfare. Peace also heralded a new era for Georgetown and the rest of the United States as well. Gone were the days of isolation. The country now had become a world leader and a member of the United Nations, and was about to become involved in a "cold war" with the communist Soviet Union. America's recent ally in the struggle against the Axis, Russia had now become the new opponent to democracy on the world scene, while an "iron curtain" had descended over much of eastern Europe.

Looking around, Americans realized that their world had changed dramatically in the past four years. The same might be said for Georgetown folk, in their isolated homes nestled along Clear Creek. Their once sole economic support, mining, could no longer be depended upon as a foundation. Something else had to be found to keep the community alive and vibrant. The people looked about and realized something would be needed to prevent Georgetown from going down that path toward "ghost town" status, which many of their once-urban neighbors had already attained. The answer lay all around them—a charming valley tucked in among strikingly beautiful mountains.

If they promoted these things, tourists would come. But did they really want that to happen? Changes would undoubtedly arrive, outsiders would come in, lifestyles would change, and their town and valley might never be the same again. The residents and community leaders confronted a fork in the road, and the way they selected to travel would chart the future.

There was nothing new in this. The English poet, Edmund Spenser, wrote 400 years before: "but times do change and move continually." But change could be both progressive and threatening. Nor could Georgetown residents assume that their fate rested solely in their own hands.

Georgetown, in the post–World War II era, struggled, like many other American small towns, trying to return to peacetime pursuits and, at the same time, finding its place in this new world. Perhaps the problem was worse for a community located in a valley penned in on three sides by mountains. The old economy, except for tourism, had faded nearly into insignificance.

While Georgetown residents pondered their future, Colorado was changing beyond their valley. People flocked to postwar Colorado, as was shown by the fact that Denver's population by the 1950s topped 450,000, a third more than it had been before the war. Front-range communities everywhere showed a similar vitality. The state's Democratic wanderings in the 1930s proved brief. Colorado overwhelmingly voted for Republican presidential candidate Dwight Eisenhower in 1952 and 1956. Ike proved particularly popular because he loved to trout-fish along the state's streams while on visits to his mother-in-law, who lived in Denver.

Life went on as usual in Georgetown, as it always had since the boom days had become yesterday's dreams. A glimpse of the town—and its residents—during these years may be seen in the columns of the *Georgetown Courier.*[1]

Hope sprang eternal in the January 10, 1952 issue. The reopening of the old mines, "which many are on tap to do at present," would be highlighted at the annual "Gold and Silver banquet." Placer mining, particularly, was making a comeback, harkening back to the earliest local mining in the days of the Pike's Peak rush. All Georgetown miners were also urged to go to the Colorado mining convention, which would be held in Denver later that month.

Local news featured the fact that the city council had voted to increase the water rate to $3 per month and, at the same time, had raised the marshal's salary to $25 per month. There was concern about waste material, fertilizer, mine tailings, and other things polluting local streams. What could be done about that? The feature story in the May 22 issue highlighted the "impressive commencement ceremony," at which four girls and two boys received high school diplomas. Three girls graduated from eighth grade. Speaking of children, a playground was being leveled flanking the Alpine Hose building for their use.

The year 1953 dawned crisp and cold over Colorado. The new Boulder-Denver turnpike was proving what its supporters had predicted it would do, as traffic increased steadily every month. While Denver's suburbs benefited from this transportation artery, these communities would continuously grow on their own during the postwar boom. Front-range towns also benefited from incoming government facilities, as the Cold War tightened its hold, and government offices moved west for security and safety reasons.

Such matters had little impact on Georgetown. Wishing to become part of the "new" Colorado, local residents proposed to organize a businessmen's club. Its goal was "to accelerate and promote a more rapid growth and development for the town."

While the town tried to stay isolated from state affairs and the increased regulations starting to come from the state legislature, several changes were coming their way. Changes to the state highway system would require annual reports on all roads within the municipality. Pollution in the creeks, which started in the high country and merged around the Denver metropolitan area, caused the cities on the plains to complain to their state legislators, who soon decided to look into the water and sewage systems along the Platte River drainage. The town's ninety-year-old infrastructure—roads, waterlines, fire equipment, and cribbing—all needed serious attention. The townspeople who had served on the various boards and commissions were getting weary to say the least. No one stepped forward to run for the post of police judge (ex officio mayor). The *Courier* jumped into the fray with an editorial. "During the next few years Georgetown is going to need a man who can lead it into an era of expansion and prosperity. Political affiliations cannot be considered. Petty difference and 'axe-grinding' must not interfere with the administration of the town affairs during the critical days ahead which will determine the future status of Georgetown."

The man who would come to the front was James Grafton Rogers, attorney and statesman. Rogers and his wife, Cora, had come to town in 1941. In May of that year Rogers had received a call in Mexico from his former boss Secretary of War Henry Stimson. Rogers had worked for Stimson as undersecretary of state in the Hoover administration. Stimson wanted to head west with General George C. Marshall, army chief of staff, to meet with Rogers among other things. Rogers was soon to resign from the Timothy Dwight College at Yale and was traveling in Mexico. He indicated that he did not have a home in Colorado at this point, but was sure they could work something out.

Later that day Rogers's son Ranger called to say he had found a house for sale in Georgetown that he thought his parents would like. They quickly purchased the home built by Robert Orchard Olds on the northwest side of town, which they soon christened "Aspen Grove." Rogers would later talk about the meeting the three men had in his newly purchased Georgetown home.[2] The house was well suited for a quiet meeting, which Rogers later described as part of a series of weeklong meetings on "army affairs."[3] The exact nature of the discussions are unknown; however, his son-in-law Stephen Hart would later note that Rogers "helped break a bureaucratic log jam which led to approval of the creation of the famed 10th Mountain Division, the elite group of highly trained mountaineering troops who learned their trade at Camp Hale in Colorado, drove up the chain of Apennines under the terrible conditions and broke into the valley of the Po."[4] As the founder and first president of the Colorado Mountain Club, and three-time president of the American Alpine Club (including a term right before his move to Georgetown), Rogers certainly knew the value of specialized mountaineering skills. The division was activated in November 1941, just a few months after the meeting in Georgetown.

Through the years, as local residents began to get to know the Rogers family, many felt that important decisions and high-level discussions were commonplace within the walls of the lovely old Victorian home. The other issue that may have been discussed during the Rogers-Stimson-Marshall meeting in 1941 was the need for a civilian intelligence gathering agency. Rogers had written a letter to President Franklin Roosevelt suggesting the formation of such an agency outside of the realm of the military. Rogers believed that the experiences of the British and German military intelligence services proved his point. Apparently Roosevelt set the letter aside without taking any action, which Rogers believed stemmed from their political differences, since Rogers had been a member of the Hoover administration. When William Donovan approached the president with a similar idea about the same time, Roosevelt gave him Rogers's letter, and discussions about a civilian intelligence agency started in earnest.[5]

Rogers moved to Washington in 1942 to help with the formation of the Office of Strategic Services (OSS) and to serve as chairman of the planning board and one of the top assistants to director "Wild Bill" Donovan. After a couple of years, Rogers retired from the OSS and moved to New York City with his wife; however, they still visited Georgetown and their Aspen Grove residence whenever possible. When he announced his pending retirement from the OSS, January 1944, Hugh Wilson sent a letter to him, in Georgetown, telling him he would be missed: "I only hope that whatever you do, in Congress I hope, if not, elsewhere, you will have something to do with the big show because I fear that you will not be content otherwise."[6] Wilson would certainly have been surprised to learn that within the next ten years Rogers' political ambitions would center upon Georgetown, Colorado, and two terms as police judge, ex officio mayor.

James and Cora Rogers had moved back to Georgetown on a full-time basis by 1953. At age seventy Rogers was ready for retirement, at least as much as would be possible for someone who still served on a variety of state and national boards and commissions. The townspeople approached Rogers, who hesitated then decided to throw his hat in the ring. "If the people of Georgetown elect me as Police Judge and by reason of that office as Mayor of the City of Georgetown I will serve gladly and to the best of my ability on behalf of the community." He continued in his letter to the *Courier*, "I have no desire to hold any more public offices, but Mrs. Rogers and I have a deep interest in the town where we live and I am willing to accept the post if you think I can be useful."

The *Courier* definitely thought he could. Rogers's background included a degree in law from Yale, along with several positions in academia and government posts that the townspeople were noticing as the family had come and gone through the years. The *Courier* supported his candidacy. Rogers received "the designation by both the Citizens and the Taxpayers' Committee, which would appear to make his election assured. Congratulations to both Mr.

Rogers and Georgetown." In later years his son-in-law, Stephen H. Hart, noted that his election was "unanimous," something not common in a small town.[7]

Rogers's first two-year term started with a variety of issues, all of which required money, one thing the town's 400 residents did not have much of. The town's fire department had hoses to attach to aged hydrants, which tapped into water lines from the 1890s, but no truck or pumper, and insurance companies were not happy about it. The town's insurance rating was about to plummet unless the town acquired a fire engine.

Rogers and the town board set out to determine what the fire department had on hand, what would be needed, and what it would cost. A steering committee was appointed with four subcommittees: equipment, constitution and bylaws, fire protection, and finance. The committees included a variety of longtime and newer residents. At the same time, the department went under a general reorganization. Steve Nelson was appointed as temporary chief (later replaced by longtime resident Doug Marshall), with Harold Haskins as secretary, and Paul Ecklund as treasurer. These men would serve the town well over the next few decades.

A new truck would cost about $6,000, and there would be need for additional equipment. There were several used trucks available, but the townspeople and the remaining twelve members of the fire department decided to set their sights on a new truck. Rogers contacted the directors of the American National Bank in Denver, who offered to finance up to 70 percent of the cost at 6 percent interest.[8] By the time the order was placed with Howe Company, the cost was $6,153. The fire department and women's auxiliary raised $3,300, and the town financed the remaining $2,853.

At the same time, the state legislature started gathering information and holding hearings regarding the quality of water in the South Platte River drainage. Towns were requested to provide information on their municipal water and sewage systems. Rogers summarized the situation in a letter to Ralph M. Stuck, a member of the Colorado House of Representatives and chairman of the Health and Welfare Committee: "We have no municipal sewer system. Most of the town (all east of Clear Creek) connects with old private sewers which the Town in practice maintains in order. Perhaps 90% of the houses occupied in winter use this sewer, say 70 houses. It flows directly into Clear Creek. Clear Creek falls rapidly and no other town lies within about five miles. We have no plans for a disposal plant and the requirement would I think paralyze the Town. Its total income, chiefly from water rents is about $8,000 per year."[9]

Needless to say, the response was not what the state representative wanted to hear. The state continued to write letters to the mayor pointing out elevated levels of pollutants, but the town managed to avoid the issue. The state issued many stern warnings, but took no action. Apparently there were bigger issues with other towns along the South Platte drainage.

The town had more immediate concerns. During runoff season June 10, 1952, a twenty-foot gap had opened up on the western edge of the old White Dam north of town. The concrete-and-log structure continued to hold, but it seemed to be just a matter of time before the dam would break completely. The dam's breach had sent a torrent of water north and east through the length of the canyon, taking out the westbound lane of US 6 and 40 east of Idaho Springs, as well as washing out the bridge below the dam north of Georgetown. The account in the *Denver Post* noted that the new road (US 6) running west of Golden to the forks of the creek near the corner of Clear Creek, Jefferson and Gilpin Counties, had not been damaged.[10]

About the same time, the owner of the Lupton, or "Commonwealth," Mill wanted to release tailings into the lake, which would have attracted the attention of the new "pollution" committee in Denver, as well as others with property below the dam. Rogers drew up injunction papers in February 1954. Those who felt mining could still be the basis of the town's economy were livid, and undoubtedly questioned their support of the "new" mayor.[11]

Amid these major problems, Rogers found himself penning letters to town residents reminding them that they needed to pay their water bills or face losing service. Many of the town's residents had stopped paying property taxes and water bills during the Depression as they struggled to survive. Nevertheless, the town needed money to take care of existing water lines and needed to start making money to pay for the anticipated state mandates regarding sewage.

While Rogers was signing proclamations about town cleanup and painting days, at least one resident, Bernard Buckley, had been allowing his horse to roam the town, destroying yards and eating gardens. Rogers took his pen in hand September 30, 1953:

> For many months you have allowed a white horse which is your property to remain at large in the Town limits. This is an offense forbidden by Town law

> since its early days and punishable by fine and imprisonment if the fine is not at once paid. Your fellow citizens have complained repeatedly to the authorities. Your horse injures property and endangers children and threatens injury on the public highways. . . . It is insecure, illegal and even cruel. The Marshall has warned you repeatedly for months.
>
> If this animal is found at large again in the Town, the Marshall is instructed to arrest and bring you for trial before the Police Judge.[12]

Six months into the job, Rogers may have been longing for the good old days when his projects were easier—something like creating a national intelligence agency, solving the German war debt, or completing the St. Lawrence Seaway—all of which he had done in the past, undoubtedly with less aggravation.

Nevertheless, Rogers wanted to find ways to help the town develop a business base other than mining. He found a kindred spirit in Denver businessman, inventor and engineer, Marcellus Merrill. Merrill had been coming to Georgetown for years in search of a quiet place to work on projects. He had used the Hotel De Paris as a retreat during this time. Merrill's patent of a machine to balance car wheels had attracted the attention of and contracts with several large automobile companies in Detroit. As the army started to consider the establishment of an air force, Merrill was contacted to see if he would be able to work on a system to balance and coordinate propellers for the Department of the Army. Merrill indicated that he was interested, but would require a "C" coupon book to allow him to purchase gas to travel to Georgetown, where he could get away and think without the distraction of his business or young family. He got his coupon book and commenced work on more inventions. His mountain sojourns were successful, and he soon began to bring his family up to town for jeep trips, skiing, fishing, and other adventures. In 1945, the Merrill family purchased an old miner's cottage at the corner of Eleventh and Main Streets, and soon became well known and respected as part-time residents.

Merrill and Rogers, brilliant men with distinctly different backgrounds, became friends and started to discuss ways in which the town could develop a tax base or business plan that might help the town move from its dependence on mining. Rogers hoped that Merrill's knowledge of business could help the struggling town. In March 1954 Rogers wrote to Merrill about the idea of a town industry, noting "the town has now no industrial activity, as the Lupton enterprise has collapsed and the Ohlander [machine] shop is having problems . . . I cannot yet see a practicable program. Please let me have all the suggestions you have." Merrill responded a week later: "When, though, you ask me for practical suggestions of how we could get more industry into the small town that it needs so badly on that I will have to admit I am stumped . . . [and] that most ventures would require and also would demand the management devote unselfishly a good deal of time for several years without much financial return. These individuals, due to the present high cost of living, are getting increasingly hard to find." Later that year, Merrill would suggest that "Williamsburg is a criterion for Georgetown if we would just follow its example. Of course we do not have a Rockefeller to back us. We could do though something at this time. We could have the town board pass on the architecture of any new buildings to see that they fit in with our quaint atmosphere. . . . If we move now, Mayor, Georgetown will find itself in a very enviable position in a few years. It might be possible you would find people travelling to visit Georgetown as they now do Williamsburg."[13]

Rogers responded with a letter summarizing both his vision for the town, and the concerns of some of the townspeople:

> I was glad to get your letter in regard to zoning the Town and have shown it to several people. I have been talking about zoning for many months, to the Board of Selectmen and the people. I think the Board, originally adverse is favorable now but the smaller business people seem to feel that anything which puts any check on any sort of development in the town would hurt them, that the move to preserve the town's present form is in the interest only of the well-to-do who do not depend on general traffic. [Town Attorney] Henry Luedde contends we have no power to zone. He thinks the state statute authorizing towns to zone does not extend to our special charter. This is a favorite legal position of his. I think it erroneous in this case.
>
> I am not sure the Town will support a zoning ordinance after my term expires in the Spring. I am confident that anything we can do to preserve the village and even restore its early form will be wise policy and serve everyone's interest, even the small business people but this view takes some experience.[14]

While many of the new people, or summer residents, delighted in the old town, many of those who had lived in the town as it continued to shrink in size and lose jobs as the mines closed, wouldn't have shared the vision of these two men.

About this time, however, one advancement was greatly appreciated by the residents of Georgetown. In 1954 local resident Melvin Reichwein brought cable TV to town residents. Reichwein had served as an officer in the Army Signal Corps, returning to town with an honorable discharge in 1946. Two years after television finally made it to Denver in 1952, he decided to place a receiver in the vicinity of Highland Park, just over the ridgeline east of town. The receiver was connected to Georgetown homes through a cable running down the face of Alpine Mountain. Reichwein's wife, Rosemary, recalled how much the local boys were intrigued with the operation and wanted to travel up the mountain with Mel anytime there was a need to check the operation.[15] The small, volunteer business gradually grew, and Mel finally had to start charging people. The town issued a franchise license to Mel's T. V. and Appliance in 1965, which required the licensee to guarantee reception of channels 2, 4, 7, and 9. In 1967 Reichwein sold his cable interests, which included approximately 150 hookups to Glenn W. Jones, who secured part of the down payment by taking out a loan on his much-beloved Volkswagen. After his start in Georgetown, Jones would move on to develop a franchise with over 1.4 million viewers, known throughout the country as Jones Intercable.

In 1954 Rogers completed the questionnaire sent by the Colorado State Planning Commission for the "Colorado Year Book." He listed the population as 350 in the winter and 800 in the summer. There was one-half mile of paved road, three miles of graveled road, and five miles of unsurfaced road. There were no hotels, 2 motor courts with a total of 12 rooms and 3 guest houses, making a total of 20 rooms available to visitors. The town had approximately 175 telephones and 200 electric meters. Under recreational facilities, Rogers listed one park, "many" ski areas nearby, and "fishing, climbing, museums, skiing, mountaineering." Under remarks, he noted, "It is clearly one of the two or three most beautifully situated and quaintest of Rocky Mountain towns. Two museums are open. Hotel De Paris Museum (owned by Colonial Dames Society Historical Foundation) and Hamill House, privately owned: both patronized by thousands every year. Two large streams with good fishing and a lake are within the town: innumerable recreation points in environs. Probably 50,000 tourists stop in town every year."[16]

Rogers took special delight with the opening of the Hotel De Paris Museum in 1954. As president and later chairman of the board of the Colorado Historical Society, a post he would hold until his death in 1971, Rogers had studied the state and worked on developing a plan for regional properties and museums. Funds were limited, however, making acquisitions of new properties difficult. When the National Society of Colonial Dames of America in the State of Colorado, approached him in 1953 asking for ideas about a project for their organization, Rogers brought up the idea of the Hotel De Paris.

Chapter president Miriam W. Adams later recalled a meeting with Dr. Rogers in November 1953 where he spoke of the hotel and the opportunity to purchase and save the wonderful old building. Mrs. Adams said the organization had always thought they might want to own a property "some day," and decided to put in an offer of $13,000, which she felt was a leap of faith in that the organization had about $50 of unrestricted cash on hand at that moment. The hotel was owned by Hazel McAdams, whose mother, Sarah Burkholder, had always hoped it could be a museum.

The purchase was completed in time for Rogers to include the museum in his 1954 report to the state planning commission along with the optimistic prediction of thousands of visitors. The building would be lovingly maintained and restored by the women of the Colonial Dames. Mrs. Adams would thank Rogers for his vision and help with the project.[17] It is unknown if Rogers wife, Cora, had much to do with the acquisition, since she was a member of the Colonial Dames. Her mother, Lucy Peabody, wife of Governor James Peabody, had been instrumental in efforts to protect Mesa Verde, with some referring to her as the "Mother of Mesa Verde." Work with historic properties may have been in Cora's blood.

In 1955, near the end of Rogers's first term as police judge, the State Game and Fish Department agreed to repair the dam at the north end of town, using the existing dam as the basis for the new structure. Unfortunately, the quality of the work would be questioned on June 4, 1956, when the new spillway broke. Preliminary damage assessments were placed at $400,000 with twelve bridges washed out and several stretches of road severely damaged, including 200 feet of the main highway west of Idaho Springs. Approximately thirty families in Clear Creek County were stranded on the southern side of the creek. The *Courier* reported: "The first sign of trouble starting at the spillway was seen by Bill Menzies . . . [who] noticed that water below the spillway was muddy but the water going over the spillway was clear. . . . A spiral of water, such as forms above the drain in a sink, was growing on the lake side of the spillway and more and more muddy water was coloring the lower side of the spillway."[18]

Menzies notified Sheriff Jim Sacra, who contacted County Road Supervisor Byron Guanella. A state patrolman was passing by at the same time. Guanella rushed down the road ahead of the flood, notifying residents of the pending danger, undoubtedly saving many lives.

The dam gave out with water rushing under the spillway wall. All that was left was a structure about 4 feet high and 100 feet long. As the county towns and residents began to assess their damage, questions arose regarding liability. The *Courier* noted: "Rogers Says Town Not Liable For Damages," nonetheless, the fact that the structure hadn't lasted through its first runoff season left people pointing fingers in all directions.

Lieutenant Governor Stephen McNichols declared a state of emergency, and contacted the federal government with a request for aid, noting that he had been in contact with Governor Ed Johnson, who was in Washington discussing the proposed interstate highway system. President Dwight Eisenhower signed the designation of a national disaster area along with authorization of $50,000 toward repairs from his bed at Walter Reed hospital in Washington, DC, where he was recovering from intestinal surgery. Presidential Press Secretary James Hagerty noted that the president "'read all the papers through' on the Colorado flood and that additionally 'he knew the area personally and knew of the flood conditions.'"[19]

As the estimated cost of repairs to roads, bridges, and even some of the Idaho Springs sewer system skyrocketed, Idaho Springs mayor J. Price Briscoe let his opinions be known: "Georgetown has no money. And it's not their responsibility, anyway—they didn't build the dam."[20] Briscoe was a well-known man in state politics, having served as a state senator and as director of State Institutions under former governor Dan Thornton. Certainly his voice stirred the waters as the need to pay for damages became apparent. State attorney general Duke Dunbar had spoken up shortly after the break and "insisted there can be no question of state responsibility . . . the state's immunity from responsibility prevails unless it is expressly waived."[21]

The front-page headline of the *Rocky Mountain News* the next day quieted those who were attempting to keep the state out of the affray: "State Engineer Says Clear Creek Flood Was Preventable." The state legislature would later move to waive the state's immunity and pay for most of the damages caused by the flood. At a press conference a week later, Mayor Briscoe gave a good summary of the feelings around Idaho Springs: "We definitely don't want a new lake up in Georgetown just for a bunch of lousy fish." Although he would later temper that with a caveat: "unless we have some assurance that the thing won't wash out again."[22]

The dry lake bed at the north end of town quickly became an eyesore that gave the community a "washed out" appearance to say the least. The town began discussions with the state over construction of a new dam; however, the state was very hesitant. It would take another fifteen years for the state of Colorado to decide to try again.

Rogers served a second term as police judge (ex officio mayor) between 1957 and 1959. One of the final acts taken by the Board of Selectmen in April 1959 was passage of Ordinance 131, which established a Planning Commission for the purpose of drawing up a Zoning Code to present to the town board. Rogers had noted that things were starting to change in 1955 with a letter to Marcellus Merrill:

> It seems to me the town is slowly awaking to the fact that its future is in something else than a revival of mining. The Civic Association the other night discussed earnestly raising a fund to rehabilitate some of the old structures like the fire houses they wanted to wreck last year. There was some recognition that the future of their property and their school depended a lot on dreamy stuff like history, sentiment, beauty and scenery. The transition from the decrepit mining town fossil to a residence community for people incapable of driving a drill is a big trestle to cross. For the first time I saw some daylight. The people are beginning to think that a big tunnel and a big road up Clear Creek might not solve all the future. They see that the only people who buy groceries or houses here are those attracted by the dreamy factors. I am sure that is the case. We are like the girl who took in washing only to discover that she had a fortune in her chestnut curls.[23]

The "big tunnel and big road," the construction of Interstate 70, and the Eisenhower tunnel, would loom large over the town in the 1960s, however, the town was beginning to see a new future based upon its past.

PHOTO ESSAY 10

# MODERN TIMES AND NEW ADVENTURES

*James Grafton Rogers lighting his pipe at his home in Georgetown: scholar, statesman, integral member of the team creating the OSS during World War II and police judge of Georgetown. One has to wonder which job he found the most difficult? Image courtesy of Historic Colorado (Rogers collection 10039767).*

*Mayor James Grafton Rogers grabs a can of paint to start to liven up Georgetown's buildings in the mid-1950s. The Red Ram sign in the background indicates the town's economy is starting to revive with the ski industry. Image courtesy of History Colorado f33694.*

*Georgetown taken from Strousse Park looking at the Fish Block, probably late 1940s or early 1950s. Bradley collection.*

*Once the pride of Georgetown and its owner William A. Hamill, the buildings were showing their age for this photo in 1951. The stable building in the left of the photo shows a gaping hole next to the large door. Marcellus Merrill photo, Marcellus Merrill collection.*

*This photo of Georgetown, looking southwest toward Silver Plume, was taken by Marcellus Merrill about 1950. The trees around City Park, planted in the 1890s, flourished through the years. Many vacant lots, however, showed places where buildings had been torn down, or fallen down on their own. Marcellus Merrill photo, Marcellus Merrill collection.*

*The ruins of the Catholic church burned in 1917 remain in the foreground of this photo c. 1950s. Rutherford family collection.*

*From Ben Draper's scrapbook: "This picture was taken by me in 1946 to use in an effort to get Denver antique and gift shops to buy these little buildings and open summer branches of their businesses. The two on the right have since been torn down a/c fire hazards." Ben Draper photo.*

*Buff Rutherford poses on horseback outside of the movie theater in the McClellan Hall on Sixth Street as part of the publicity for the "Big Do," in 1946. Thursday's feature was the Yellow Rose of Texas. Denver Public Library, Western History Collection x1028.*

*The Monti Block, at the corner of Sixth and Rose Streets, was one of the earliest commercial buildings in Georgetown, but by 1946 it looked as if it wasn't going to make it much longer. Ben Draper, as a partner in Georgetown Enterprises, purchased the building, started renovation, and opened the Grubstake Restaurant in the corner part of the building. Ben Draper photograph.*

*A new coat of paint, and the restoration of the Cresto & Valentine sign: "The Worst Wines, Liquors & Cigars" made the building a much-photographed landmark. Marcellus Merrill photo, Marcellus Merrill collection c. 1950.*

*A coat of new paint and a few new windows in 1946 seemed to add a spark to downtown Georgetown. The Signs in the large windows announce the anticipated opening of "The Grubstake." Ben Draper photo.*

*Ben Draper and Georgetown Enterprises also purchased the building on the southeast corner of Rose and Seventh Streets, fixed it up, and opened the St. James Bar and Grill. Ben Draper photo.*

*The smiling gent in the white shirt, behind the bar, is Ben Draper, perhaps the earliest "preservationist" and historian in Georgetown. His efforts in the 1940s, into the 1950s, would struggle financially, but everyone certainly had a good time. His projects may have been too early, but he helped to save several buildings from demolition. Bradley collection.*

*From the Ben Draper Scrapbooks: "Loading up in Georgetown for the twelve-mile trip to Waldorf where the mountain naming ceremony was held. August 1, 1948." Ben Draper photo.*

*Sixth Street was packed with cars in 1946 for the start of the Silver Stampede. From Ben Draper's scrapbooks: "This week-long celebration brought about 20,000 visitors to Georgetown (estimate of State Highway patrol). It featured two gambling casinos and a nightly "Big DO" in McClellan Hall. There were daily event such as hard rock drilling contests, horse races, foot races, etc." Photo By Ben Draper.*

*Ben Draper, in western shirt standing to the right of the flag, acted as the master of ceremony for the event held in Waldorf naming Mount Wilcox after the man who developed the Argentine Central Railroad. The dedication took place on Colorado Day, August 1, 1948. From left to right: Mrs. J. J. White, former governor John Vivian, Mary Elizabeth Wilcox (granddaughter of E. J. Wilcox), Meredith F. Buvill, executive secretary of US Branch of Geographic Names, Helen B. Wilcox, Edward B. Wilcox (daughter and son of E. J. Wilcox), and Ben Draper. Denver Public Library, Western History Collection f45615.*

FACING PAGE. *It would be hard to imagine more signs in one photo. This view of Sixth and Rose Streets was taken on September 2, 1946, as part of the Silver Stampede (the sign on the side of the building "Build the Bridge, Silver Stampede"). Proceeds from the event were earmarked for repairs to a bridge below town. The Georgetown Civic Association members served hot dogs and soft drinks. A Reno club and pool hall, advertised, and the corner of the two streets show a stop sign, directions to US 6, Green and Clear lakes, and Loveland Pass. Ben Draper photo.*

*Road, campgrounds, fishing lakes, and scenic roads all helped Georgetown survive while the town's economy transitioned from mining to tourism in the middle of the twentieth century. Clear Creek County, and road supervisor Byron Guanella, designed and constructed an automobile road (to replace a network of narrow, steep wagon roads) across the pass between Georgetown and Grant. The road would be named Guanella Pass in honor of Byron's knowledge and work on the project. Byron Guanella photo, Guanella and Buckland Family Collection.*

*William Rutherford formed a business under the name of "Colorado Crafts." Marcellus and Geraldine Merrill take the reins of the company wagon in an effort to promote Colorado crafts during one of the town events. Marcellus Merrill collection.*

*The weeklong events of the town's "Big Do's" started to overwhelm the local merchants and volunteers; however, the Georgetown Civic Association continued to sponsor dances, talks, and other events as shown on this poster from August 1951. Marcellus Merrill collection.*

THE McCLELLAN PLAYERS

Georgetown, Colorado

Proudly Present

Their Second Great Show of the 1959 Season

FIREMAN, SAVE MY CHILD!

FRIDAYS and SATURDAYS at 8:00 P. M.

August 8 thru September 5

IN THE McCLELLAN TENTS

CAST OF CHARACTERS

MRS. JULIA QUINGLE ........ Rhoda Abramoff
(Was ever a mother so torn?)

MYRTLE JONES ........ Barbara Clark
(Loyal thru all)

SALLY CROSSBURY ........ Polly Chandler
(Of an inquiring nature!)

ARCHIBALD QUINGLE ........ George Gaubatz
(Smooth as oil)

DAISY DORRANCE ........ Jessie Lewis
(Her faith is rewarded)

BING BARNES ........ Buzz Chandler
(Hold everything!)

CHESTER QUINGLE ........ Richard Tisdel
(Was ever a brother so betrayed?)

— SCENE —

The scene is laid in Mrs Quingle's living room, in a small New England town

THE MASTER OF CEREMONIES ........ Les Sims
Before the Performance and Between the Acts

COMEDY - SONGS - DANCING

Admission: Adults $1.35 Children 90c

RESERVATIONS: Denver ALpine 5-5066 Georgetown 2751
Reservations will be held only until 15 minutes before curtain time

*Ben Draper sponsored melodramas under the name of the "McClellan Players" using talent from the University of Denver; however, local thespians picked up the name in later years, bringing local talent to the front. This program from the McClellan Players in 1959 lists the cast for "Fireman, Save My Child." Marcellus Merrill collection.*

## CHAPTER 11

# A TURN TOWARD GROWTH

The state of Colorado grew steadily through the twentieth century, from a respectable 194,327 in 1880, to 1,753,947 as of the 1960 census. The population of Clear Creek County and its four incorporated municipalities, however, had declined rapidly since that same 1880 census. With the exception of 1940, which showed a small increase, the county shrank with surprising regularity. From a high of 7,823 in 1880, the county had faded to 2,793 in 1960. Georgetown had gone from 3,294 to 307 in the same time span. Idaho Springs and Empire had both shown small increases with renewed interest in gold mining around the turn of the century, and Idaho Springs decreased more slowly than the silver towns to the west, however, it dropped from a high of 2,205 in 1900 to a low of 1,480 in 1960. Empire peaked at 279 in 1900, shrinking to 110 in 1960, with Silver Plume dropping from a high of an estimated 1,500 in the 1880s to a low of 86 in 1960.

During the decade of the 1960s, however, the tide began to turn. Rogers's report to the state in 1954 counted 350 residents in the winter and 800 in the summer. The subsequent decade would start to bring winter residents to town, many of whom would be employed by the burgeoning ski industry and the restaurant and lodging businesses developing to accommodate those rushing to the slopes.

As the country shook off the shadow of the Second World War, perhaps no industry captured the joy of life without fuel rationing more than the culture of skiing in the 1950s and 1960s. Young people flocked to the mountains, some with jobs, others willing to "ski bum" or work as highly educated maintenance staff as long as they could get a free pass and ski to their heart's content. The industry had begun to develop in Colorado in the 1930s, yet had been largely closed down due to the war and the shortage of supplies, particularly gasoline. From the middle of the 1950s into the 1960s, however, the sport became wildly popular with beginners from all walks of life coming to the mountains along with accomplished skiers from the northeastern United States and Europe. The rush to the Rockies brought memories of the 1860s when the children of northeastern families headed west while their parents shook their heads wondering why their offspring were attracted to mountains thousands of miles away.

While ski towns such as Steamboat and Aspen had their own clientele, Georgetown was in the perfect location to provide permanent housing as well as overnight accommodations, restaurants, and bars for the pending ski boom in the central Rocky Mountains.

The first four major ski areas associated with the population of Denver were Berthoud, West Portal (known as Winter Park after 1939), Loveland, and Arapaho Basin. While the ski train became a popular means of access for Winter Park, those driving to these early ski areas came up US 6 and 40 (later Interstate 70) to Empire Junction two miles north of Georgetown. Berthoud and Winter Park skiers followed US 40 through Empire and over Berthoud Pass, while those headed for Loveland and Arapaho Basin continued west on US 6. Even though Georgetown had shrunk to a ghost of its former self, it was still the largest town in the western end of the county and quickly became a focal point for those who worked at Loveland and Arapaho. Two vans, blue for Loveland and green for Arapaho Basin, drove around town mornings during the winter months picking up employees for their respective ski areas.

Many early ski pioneers such as Thor Groswold, founder of the Groswold Ski Company, were located in the Denver area. Part-time Georgetown resident, Marcellus Merrill worked with Groswold on putting metal edges onto downhill skis. Merrill also developed the three-prong binding for cross-country skis and an early ski rack system for cars.

Three former members of the Tenth Mountain Division moved to Georgetown during this time period to help develop the ski industry in Colorado: Pete Seibert, Larry Jump, and Gordon Wren. Seibert took over management of Loveland Basin in 1955 after a group of investors purchased the Loveland Ski Tow Company, Inc. from Allen Bennett, establishing the Clear Creek Skiing Corporation. The business partners of the new company immediately began to develop the area from a smaller "mom and pop" operation to a true ski area. Seibert had been severely injured in the war and told he would never walk, much less ski, again. His hard work and dedication proved his doctors wrong. He managed Loveland Basin for a couple of years, but his true love would be the mountains and valley that would become known as Vail. Earl Eaton, another Georgetown resident at this time, had surveyed a promising mountain range in Eagle County. Eaton and Seibert climbed the range in 1957, and decided they had found the perfect spot for their dream resort. After Seibert moved to the western slope, Roy Parker came over from Aspen to manage Loveland for four years, including a period of growth related to interest in the winter Olympics at Squaw Valley, California, in 1960. Gordon Wren of Steamboat Springs followed Parker and brought the area national recognition based upon his prowess in all aspects of competitive skiing. Wren had served in the Tenth Mountain Division, later returning to Colorado, where he became known as a competitor in both Alpine and Nordic ski events. During the 1948 Olympics he had qualified for all four categories of skiing: alpine, cross-country, ski jumping, and Nordic combined. He had been named to the National Ski Hall of Fame in 1958 and brought his well-recognized name to Loveland Basin.

In 1964 Otto Werlin took over the management of Loveland Basin, inheriting a midsize ski area with two lifts and one rope tow. During his twenty-eight years as manager, Werlin would be known for his business acumen and managerial skill. He added five chairlifts, and numerous trails including the development of the adjacent Loveland Valley area. Werlin worked with area owner Chet Upham (a Texan with a solid background in engineering who loved Colorado skiing and put years of time, effort, and funding into the Loveland ski area) to develop snow-making facilities in the late 1960s, tapping into the considerable power being used to build the Eisenhower Tunnel right next door to the area, a major improvement for the Colorado ski industry.

Another Tenth Mountain veteran, Larry Jump, moved to Georgetown in the 1940s. In 1945 and 1946 Jump had been hired by the Winter Sports Committee of the Denver Chamber of Commerce to scout potential sites for ski area development. Jump and Frederick "Sandy" Schauffler recommended the west side of Loveland Pass, and soon decided to try to build the area on their own. Financial problems quickly reared their head. In 1947 the area had one rope tow, $45,000 in debt, and $117 in the bank. About that time, young Marjorie (Marnie) Brown, a graduate of Bennington College in Vermont, and former member of the Women's Army Corps (WAC), moved to Colorado with an interest in skiing. The mountains around Arapaho Basin, and manager Larry Jump both caught her eye. After looking over the dismal financial situation, however, she decided to use her notable business acumen, and connections with east coast families to help the struggling area and earn herself a job as vice president of Arapaho Basin Corporation.[1]

Larry and Marnie married and made Georgetown their home. The town continued to gather an impressive number of skiers destined to change the growing industry. German native Willy Schaeffler moved to Colorado in 1946 and went to work for the Jumps directing the new ski school at Arapaho Basin while living in Georgetown. Two years later Schaeffler would become the coach of the University of Denver Ski Team, a position he held for many years.

Old houses around town found new owners. The Buff Ski Club of the University of Colorado had purchased a building on Eighth Street prior to the war, which soon filled with young skiers throughout the winter. The Schussbaumer Ski Club purchased three small buildings on Brownell Street in 1952, which they developed into a ski chalet with kitchen, bar, living room, and dorms.[2] Other skiers needed places to stay, as well as restaurants and bars for the new "sport" of après ski. The Red Ram Restaurant opened by William Holmes and his wife, Annette, became the place to stop for drinks and German food. Patrons were advised not to dance in their ski boots.

A new motel, later known as the Georgetown Motor Inn, was the first prefabricated building to be built in town. Owners Ralph Ennis and Neil Williams had the building delivered and put together in less than a week.[3] Georgetown's restaurants and bars started to attract musicians as well as skiers, with folk singer Harry Tuft performing at the Holy Cat, and Walt

Connelly performing at the Red Ram. Dick Kroeckel and his piano became another house favorite at both the Ram and the Silver Queen.

As the number of residents and visitors increased, special events started to develop. The most enthusiastic event promoter had been Ben Draper and his company: Georgetown Enterprises. After mustering out of the coast guard after the war, Draper had returned to Colorado with his love of history and theater. The enthusiastic young man had spent time working with Ida Kruse MacFarlane and Anne Evans after they restored and opened the historic Central City Opera House in the 1932. Draper had secured a PhD in economics from the University of Chicago, and would later complete another PhD in theater at the University of Denver, with a focus on the history of Colorado theater. He decided to follow his two favorite academic pursuits in Georgetown, where he hoped to find a way to restore the town's buildings with funds from his various enterprises.

His company, Georgetown Enterprises, was created in 1945 as a partnership with Jack Simmons of Denver. Draper had purchased a house at 919 Taos Street in 1930 for the whopping sum of $30.50 in back taxes; however, he was generally considered a Denver resident. He would later explain the 1940s project as follows: "A commercial venture, the aim of the firm is to restore Georgetown to its 1880 glory both as a tourist attraction and a thriving community."[4] Draper purchased the west half of the block at the northwest corner of Sixth and Rose, from Mrs. Maria Cresto in 1945. He proceeded to open two restaurants: the St. James and the Grubstake. At the same time, he purchased the little McClellan Hall on Sixth Street (not to be confused with the McClellan Opera House, which had burned to the ground in 1892); he used it for a movie theater, melodramas, and other events.

The young man began to send press releases to papers throughout the country. In 1946 the *San Francisco Examiner* noted:

> Georgetown, high in the Rocky Mountains, is being restored as the famous mining camp it was in the 1870's under a five year plan adopted here.
>
> The restoration, directed by Benjamin Draper and Jack Simmons, both World War II veterans, will provide a modern example of mining town architecture, even to the color of the paint.[5]

A headline in the *Rocky Mountain News* proclaimed: "New Life Floods Georgetown's Veins; Town Preparing Royal Welcome To Tourist Army."[6]

As might be expected, the locals began to worry about the arrival of this "army" of tourists and the potential impact on their quiet lifestyle. The *Denver Post* announced: "Georgetown Plans Mass Meeting for Program Discussion." The two-paragraph article announced a meeting sponsored by the Georgetown Civic Association to hear concerns from the citizens about "a problem common to many small Colorado tourist centers—the conflict of interests between townspeople and summer visitors." Proposed projects mentioned in the article included "a highway advertising campaign, oiling of residential streets, stocking fishing streams, a scenic highway from the top of Mount Evans to Georgetown, new street lights and erection of a commemorative tablet on the site where Louis Vasquez built his cabin in 1832." As if that weren't enough to ruffle a few local feathers, the article ended with the following request: "Property owners will be urged to provide living accommodations in their homes, if possible, for visitors."[7] The enthusiasm of the men of Georgetown Enterprises was definitely starting to wake up the sleepy little town.

Work started almost immediately on the restoration of the Monti Block, which was in serious need of a coat of paint and included touching up the interesting sign on the west side of the building: "Cresto & Valentine The Worst Wines Liquors and Cigars." Once the paintbrushes were out, Draper moved to repainting the US 6 sign at Empire junction with Delmar "Squeak" Morris and Craig Abrahamson.[8] On Saturday, July 13, 1946, attendees of the Sixth Annual Western Folklore Conference being held at the University of Denver were offered a trip to Central City and Georgetown, the start of several years of tours for groups gathered for meetings in the metro area.[9]

Perhaps the biggest event of 1946 in Georgetown was aptly named "The Big Do." Draper used his ties with the University of Denver to hire the D. U. Promenaders (square dancers) to spend the summer in town putting on performances every Saturday night in the McClellan Hall building. The dancers were to be housed in town and fed at the Grubstake Restaurant. About the same time, the movie theater opened with Bing Crosby and Ingrid Bergman in *The Bells of St. Mary's*. Aware of the need to make sure the local residents were included, Draper gave a party for "Georgetown folks" on June 3, 1946, a few nights before the opening event for the general public.

Opening night for the Big Do was June 8, 1946, with a sellout crowd at the Grubstake for a $2.00 dinner, which included "fresh vegetable soup or tomato Juice; lettuce and tomato salad with mayonnaise or French dressing; fried chicken, mountain style or mountain trout; mashed or hashed brown potatoes; turnips, or buttered carrots; apple or peach pie; coffee, tea or milk." The menu noted that "all prices as low or lower than OPA [Office of Price Administration] Ceiling price" along with an apology that the liquor license had not been finalized as of the date of the opening.

For the Fourth of July, the locals were on board as the *Denver Post* noted that Henry Anderson, president of the Georgetown Civic Association, announced plans for a "Big Picnic" and fireworks. "The town has freshened up picnic spots and merchant members of the civic association are stocked to care for an unusually large number of visitors."[10]

The events continued over the next few years, including a ten-day Silver Stampede in 1946, with ten local women given the honor of serving as queen. According to the *Rocky Mountain News* the ladies held court as follows: "Mrs. Maurice Anderson, wife of Georgetown's mayor, will wear the crown for the first opening day and first of the nightly 'Big Do's' to be held in Alpine St. under the stars. Mrs. Henry Anderson, wife of the president of the Georgetown Civic Assn., will take over the crown for Sunday's celebrations. The other queens are: Mrs. Thomas Abrahamson, Aug. 26; Mrs. Mary Ann Lee, Aug. 27; Mrs. Bernard Buckley, Aug. 28; Mrs. Melvin Reichwein, Aug. 29; Miss Ruth Skeen, Aug. 30; Mrs. Louis Anderson, Aug. 31; Miss Betty Whitenack, Sept. 1, and Miss Phyllis Buckley, Sept. 2." The article went on to note that several of the children's events had been cancelled due to concerns about the current polio epidemic, but the hardworking members of the Georgetown Enterprises still intended to sponsor the horse races on Taos Street, tugs-of-war, rock drilling, and baseball games.[11]

The next day's issue of the *News* indicated that 2,500 people attended the ribbon cutting for the first day of the event. Seven of the honorary queens had their portraits published in the same issue.

Draper's involvement in special events, tourism promotion, and historic preservation marked the start of economic trends that would become central to the survival of Georgetown as mining disappeared from its mountainsides. Nevertheless, the late 1940s were not the best time for such a grand undertaking. Like the earliest investors in mining around Georgetown in the 1860s, Draper and his associates started something that would work in later years but ended up costing them much more than they ever retrieved from their investments. In *Preservation Comes of Age*, a history of the historic preservation movement between 1926 and 1949, author Charles B. Hosmer Jr. noted that Draper may have been overly optimistic in an article published in the January 1949 issue of "Antiques Magazine" in which he [Draper] explained their efforts: "The restoration of Georgetown has been through a private venture, based on the idea that the atmosphere of the past is merchantable. The backers of Georgetown Enterprises have proved beyond a doubt that in the west as in the east authentic restoration has popular appeal."[12]

Hosmer went on to note: "It had proved to be a rewarding experience although the Drapers and their friends lost heavily in the business and had to abandon Georgetown to other preservationists. But the saving of Georgetown had been based wholly on giving the town a sense of life through the love of Victorian antiquities and appreciation of the creative arts."[13] The start of preservation in Georgetown, he wrote, was in direct contrast to the "Helldorado" style of saving the Wild West in places such as Tombstone, Arizona.

In 1946 Draper made a point of photographing the town's historic structures. His love of the town's history and its old buildings may have been the start of the town viewing its history as an asset, not something that needed to be torn down and replaced. Certainly many of the town residents loved their old buildings, but over the years empty buildings had suffered through years of neglect. Several buildings had been torn down to keep them from turning into firetraps or playhouses for the town's children. Looking back, it is easy to imagine that the fresh coat of paint on the Monti Block, at the center of the town's small commercial district, was a badly needed boost to the town's image. Suddenly people started coming to town to see the old buildings—there was hope indeed.

In 1949 Draper and his wife Barbara moved to California, where he accepted a job as an executive producer with the National Broadcasting Society. The couple returned to town as time allowed, retaining their house on Taos Street, and their memories of several frantic years with Georgetown Enterprises.

The Fourth of July celebrations would continue through the years, but the "Big Do's" were replaced with smaller, one- or two-day events sponsored by the Georgetown Civic Association. In 1951, summer events started with a celebration of "Georgetown Day" on August 6, with a sports car rally and antique car parade, horseshoe pitching, fishing contest, bazaar, walking tours, music at the Presbyterian church, and a melodrama. Shortly thereafter, the "Silver Queen Jubilee" was held on August 18 and 19, with street dancing, a lecture on Colorado history by Caroline Bancroft, an auction, and African Big Game Movies by Wallace Taber of the *Denver Post*.

The Georgetown Civic Association printed flyers advertising the town, its history, and things to do. A concern was voiced, however, in a letter from James Grafton Rogers to association president Henry Anderson in 1954. Rogers asked the group to make sure they were sticking to the facts after a letter to the editor in the *Denver Post* had accused the town of false advertising. Rogers noted: "A reputation for false statements is very harmful to a community. All our experience shows this. Rival communities capitalize on such statements. There is a big body of historical students in Colorado now who are very alert about the truth. . . . There is any amount of interesting truth about Colorado and Georgetown so that even the facts are enough. The way to make Georgetown profit from its very interesting and quaint background is to build a reputation for no exaggeration."[14]

By 1956, the McClellan Players had transformed into an active group of local performers whose talents entertained residents and summer visitors alike. In 1959 *Fireman, Save My Child*, was performed in the "McClellan Tents." *Aaron Slick from Punkin Crick* was performed in the basement of the Silver Queen building. The group gradually increased the number of performances, earning the funds necessary to restore the second floor of the Star Hook and Ladder Company building across the street from the Hotel De Paris, where they would perform for years to come. By 1968, the group performed two plays, including *The Beggar's Petition, or A Father's Love and a Mother's Care!* with an Olio by "Professor" Richard Kroeckel—courtesy of the Silver Queen Restaurant.

During the same period, the townspeople continued to try special events, most of which lasted only one or two years. A change came when local merchants including silversmith Stig Gusterman and his wife, Astrid; Odette and Wally Baehler; and Bill and Annette Holmes decided to put on a Scandinavian Christmas fair (later known as Christmas Market) in December 1960. Townspeople supported the event, which had been developed to honor the Swedish traditions of residents current and past. Crafts were sold along with a variety of foods, and the local schoolchildren were taught to sing in Swedish, a tradition that continues to this day.

That same year, the townspeople started to celebrate *fasching*, a pre-Lenten event also known as *fastnacht* generally celebrated in German towns. The celebration started out small in its early years, growing into a weeklong event in the late sixties after local bartender Ronald J. Neely came on board, bringing his black tie, big ideas, and oratorical skills to the fore. In later years events ranged from the family-oriented *kinder fasching*, a parade of local schoolchildren dressed in costume, to the evening revelry at several of the local bars, and a ski race at Loveland known as the Raunch Cup. The local bars sponsored teams, with the winners bringing the trophy (a brass spittoon) back to their sponsors for the year. Skiers and townspeople celebrated the days before the start of Lent (once again, dancing in ski boots was discouraged). The town elected a prince and princess of fasching who were sent as representatives of the town to the granddaddy event in Munich, Germany. The city of Munich adopted Georgetown as a sister city, and treated the small-town visitors as guests of honor. Years of parties in Georgetown followed until the liquor laws of the late twentieth century, and the resulting concerns about liability, ended the days of carrying the prince in his coffin from bar to bar on the last day of the event. It might be said: the prince is dead, long live the prince.

In 1963 a new industry came to Georgetown with another tie to the past. Public Service Company announced plans for construction of a hydroelectric pumped storage power plant to be known as Cabin Creek, located five miles south of Georgetown on Guanella Pass. Water power had been central to the development of the mills for the town's mining industry, then helped the economy of the town through the difficult times of the early twentieth century—the small hydroelectric plant at the east end of Sixth Street had been producing power since its construction in 1900. The new plant, however, would be a state-of-the-art facility designed to allow the generation of power for addition to the regional grid. Two reservoirs would be constructed: one at approximately 11,100-foot altitude along the ridgeline of Paines Mountain, and the second just below the Guanella Pass Road, at 9,980-foot altitude. Water could be pumped to the upper reservoir during a period of low-energy use, then dropped through a penstock to the plant at the lower reservoir during peak-use periods.

Construction started in June 1964 and was completed in April 1967. The Edison Electric Institute presented the "Edison Award" to Public Service Company in recognition of the innovative technology. During the construction period, the company worked with the town to pave Rose Street, replaced the bridge across the Leavenworth branch of South Clear Creek near Tenth Street so that it could handle the heavy equipment required for the construction of the plant, and joined with Clear Creek County to pave Guanella Pass up to the Cabin Creek facility. The town finally had a "new" industry, one that could help the community with projects and employment, the kind that James Grafton Rogers and Marcellus Merrill had been searching for in the late 1940s and early 1950s.

About the same time, the new designs for Interstate 70 were being unveiled. The highway that kept the town alive, and had started to bring thousands of visitors to the town, suddenly seemed like a two-edged sword. In order to climb the grade between Georgetown and Silver Plume, the road would have to be carved into the side of Republican Mountain. The other towns in the county would be split in half by the new 200-foot right-of-way and four-lane highway. Georgetown was fortunate to escape their fate; however, the preliminary design for the big road did not include an eastbound exit lane. Suddenly, it looked as if the treasured après-ski traffic would be denied an entrance into town.

Several buildings near the entrance to town were moved to make room for the highway and the sizeable pile of cantilevered rocks, dirt, and mountain required to support the road above. James Grafton Rogers's beloved Aspen Grove house was moved to the foot of Guanella Pass. The Schussbaumer Club dismantled their ski chalet, salvaged whatever they could, and built a new lodge in Breckenridge.

The blasting required for the construction of the new road subjected the town to the same volleys of flying rocks that it had endured eighty years earlier with the blasting for the construction of the Georgetown Loop. One particularly large rock landed in the attic of the old Hamill House on Argentine Street.

The town finally wrangled an eastbound exit through the complaints of the federal engineers, who felt the exit ramp would be too steep. One serious problem remained: the protection of the railroad beds and history of the Georgetown Loop Valley. Once again, James Grafton Rogers's quiet ways and contacts with the Colorado Historical Society would find an answer.

During his tenure as President of the Colorado Historical Society (1949–59) followed by years as Chairman (1959–71) Rogers had written a "Report on the Sites and Structures of Historical Interest in the State of Colorado," in response to an act of the state legislature, which had been passed in 1953. The state's interest paralleled the work of the National Park Service staff, who had been assigned to compile "The National Survey of Historic Sites and Buildings," part of the work requested by Congress after the passage of the Historic Sites Act in 1935. The federal legislation requested a review of buildings and sites across the country in an effort to expand the role of preservation and restoration to include more than places of extraordinary beauty or battlefields. The project was delayed with the onset of the war; however, interest rekindled in the 1950s, and the Consulting Committee for the Survey and the Advisory Board on National Parks, Historic Sites, Buildings, and Monuments came up with several themes and commenced their work with the aid of historians employed by the National Park Service.[15]

Rogers' connections with Washington, DC, as well as his love of history certainly put him in a position to know of the work that was starting on the subprojects within the federal project. It would be easy to see Rogers asking the state to write the legislation to ask him to do a study that could be given to the park service historians. In any event, the report that he wrote in 1954 indicated that the state historical society had been compiling information on known sites and proposed museums.[16]

The seven-page report details a variety of sites and buildings, but ends with areas that needed further study, including railroads:

> There is no plan for a railroad museum. Our railroad history is unusually colorful. There is a large body of railroad students and enthusiasts in the State and a well established railroad club. The beginnings of an outdoor museum are made at Golden and a motel at Alamosa has gathered a considerable collection of relics. The sites most connected with notable railroad events are perhaps the Royal Gorge at Canon City, the Georgetown Loop and the Moffat Tunnel. This report deals only with historical sites or geographic points and it may be that none of these is the most appropriate place for a railroad exhibit.[17]

Rogers's opinion would change once plans for Interstate 70 were announced and the remnants of the railroad in the valley between Georgetown and Silver Plume were threatened. In 1958 Rogers spoke with Jared B. Morse, architect, about the idea of developing some kind of plan for the Loop Valley.

Morse responded in a letter with a great deal of interest: "As you will recall, our conversation on Thursday of last week discussed the qualifications of Georgetown and adjoining Clear Creek Valley for use as a restoration project, museum location, winter and summer residential development and primarily, the preservation of the once famous Georgetown Loop right-of-way."

Morse noted that he and others had been purchasing property around Georgetown "without any general program in mind" but that the more they learned about the area, the more they had become interested in its preservation. "With the preservation of the right-of-way in mind, we procurred [*sic*] the old Smith house [now known as the Pohle house] for trading purposes with the State to re-establish the north tressel [*sic*] foundation site when and if the highway is straightened. Any other useful titles that appeared were purchased. . . . We are in a position to provide some technical service for this project and would be delighted to be of any assistance to the area, and in particular, we would like to have your reaction and guidance to such a program."[18]

The project moved forward. The State Legislature awarded a $6,000 contract to Morse and Decker architects to develop a plan that included an inventory of existing structures and ideas for other buildings.[19] Rogers contacted Stanley Wallbank, who owned extensive property within the valley, and arranged for the donation of almost 100 acres of mining property to the Colorado Historical Society in 1959.

Rogers now had a project, interest on the part of the historical society, and the nucleus of a plan that could be included within the studies being done by the National Park Service and the Advisory Council. Marcellus Merrill recalled the amount of work it took to get the historic site established in a letter to the editor of the *Clear Creek Courant* in 1982: "I always think of James Grafton Rogers who spent so much time and made some 40 trips into Denver and clear back to Washington to get that park designated and get it set up. In some ways I think it should be called Rogers Park. But James Grafton Rogers said: 'A park should never be named after a man because when he passes away they'll always wonder who it was. It should be named from a natural standpoint.'"[20]

Rogers's hard work paid off on November 13, 1966, with the designation of the Georgetown / Silver Plume National Historic Landmark District, and the creation of the Georgetown Loop Historic Mining Area by the Colorado Historical Society.

The interest in the area, and pending designations, convinced the Colorado Department of Highways and the US Department of Transportation to modify the original plans to protect the Loop Valley. At one point, the agency had planned to route detour traffic up through the valley, which would have destroyed the integrity of the site. The route was changed, and the agencies became partners in the effort to preserve the history and sites of the valley. In 1969 the Colorado Department of Highways received an award from the Department of Transportation for the "Most Outstanding Example of the Preservation of Historic Sites in 1969."[21]

The interstate highway became a lifeline for the community, and skiers continued to throng to the ski slopes. Much like the nineteenth century, however, the race to the western slope brought people to Georgetown, then took them away. In the nineteenth century, railroads were hailed as the key to the economic future, but once a new railroad reached Leadville, interest in Georgetown waned and investors went elsewhere. With the construction of Interstate 70, and especially with the completion of the Eisenhower and Johnson Tunnels, Georgetown's skiers found their way to the new ski areas and resorts to the west.

PHOTO ESSAY 11

# TOURISTS RETURN TO THE VALLEY

*Marcellus Merrill, seated, with a display of his Electronic Wheel Balancing Service at the Georgetown Alpine Hill Climb, 1953 or 1954. Marcellus Merrill collection.*

*This early photo of Christmas market at Strousse Park in Georgetown was probably taken about 1965. The booths are rough hewn but did provide some shelter from snow if needed. Image courtesy of History Colorado f33693.*

*Twentieth-century miners referred to this mill as the Centennial (until the new mill was built on Third Street) or the Hazelton Santiago. It burned to the ground July 1, 1960. Photo by Anne Carman Farrow. Carman Farrow family collection.*

*The Cabin Creek pumped storage facility on Guanella Pass south of Georgetown, would provide power to a number of front-range cities. It also provided a number of jobs to residents of Georgetown. Photo from Dorothy Niles scrapbook, Rutherford family photo.*

*This early map of the Loveland Basin Ski Area shows the area as it begins to expand along the mountains to the south and east. The legend refers to 10—DownHill Race course—and 11—Switchback Trail—both shown with access from the upper switchback of Loveland Pass. Most of area included in these runs would become the nucleus of Loveland Valley, east of the basin. The Timberline and South trails show the expansion onto the upper areas of the southwest side of the Basin. Courtesy of Clear Creek Skiing Corporation.*

*Chet and Virginia Upham (left) along with Arial and Bob Murri (right). Upham and Murri were two of the original partners who purchased Loveland Ski Tow, Inc. in 1955. Others included original owner Al Bennett, Bill Bolin, and Pricilla Barnard. The Upham family has now been involved with the area for close to sixty years. Mr. Upham's knowledge of engineering, particularly piping, became invaluable as he and manager Otto Werlin installed and operated the first snow-making equipment at a major ski area in the West. Photo courtesy of Clear Creek Skiing Corporation.*

PREVIOUS PAGE. *Gordon Wren shows off some of his well-known skiing, and jumping, skills as he blasts over three seemingly unconcerned friends. From left to right: Harold "Harpo" Marks, Willy Knight, and unidentified friend. Wren, a native of Steamboat Springs, Colorado, managed the Loveland Basin Ski Area before moving on to Jackson Hole, Wyoming, then returning to his home in Steamboat to manage Mount Werner. Wren was one of the many veterans of the Tenth Mountain Division who would return from World War II and turn to skiing in the West. Photo courtesy of Clear Creek Skiing Corporation.*

*Otto Werlin's twenty-eight year tenure as manager of Loveland Basin included the addition of three new chairlifts, opening of the smaller Loveland Valley ski area immediately to the east, and the installation of state-of-the-art snow-making equipment. Many would say that he combined his love of the sport with a solid business acumen, allowing the area to grow exponentially. Photo courtesy of Clear Creek Skiing Corporation.*

*Buff Rutherford adjusting a Larchmont Snow Gun at Loveland Basin in 1967. Loveland would be one of the first Colorado ski areas to successfully use snow-making equipment, which allowed them to open weeks before other areas. Rutherford family photo.*

*Buff Rutherford and son Bruce pose for a photo at Loveland Basin in 1965. Rutherford family photo.*

*Bob Gibbs, longtime manager of food services at Loveland, came west with a love of the outdoors and an interest in business. He would later serve as chairman of the Georgetown Planning Commission and longtime member of the Board of Directors of Historic Georgetown, Inc. The photo above shows him taking a break from the Loveland Ski area to try out the slopes at Winter Park in 1962. Gibbs family collection.*

*Bob Gibbs, with sons Mike and Tim, at Loveland Valley. Gibbs family collection.*

*Larry and Marnie Jump seen dancing at their wedding. The young couple and their dedication to the development of Arapaho Basin ski area on the southern side of Loveland Pass, would have an impact on Georgetown, as many of the employees, as well as the young couple, lived in Georgetown. Jump family collection.*

*Larry and Marnie Jump at A-Basin. The Jumps became involved with the efforts to promote Georgetown and local skiing throughout the 1950s and 1960s. Jump family collection.*

*Henry Anderson cuts a steak at Kneisel and Anderson store in the 1960s. The store is now run by the fifth generation of the Kneisel and Anderson families. Image courtesy of History Colorado, Denver Post collection f33695.*

*James Grafton Rogers pauses for a photograph outside of his home "Aspen Grove," in the northwest corner of Georgetown. The trees and gardens of the lovely home would be plowed over for the construction of Interstate 70. The home would be moved to the east end of Second Street, where it would later be restored. Marcellus Merrill photo, Marcellus Merrill collection.*

*The old miners' club met at the Alpine Inn in Georgetown for a number of years. Tall tales and memories of mining in the good old days kept the group (and those who could listen in) well entertained. This photo was taken in the 1960s. From left to right: Marcellus Merrill, Price Briscoe, D. A. Knight, Walt Anderson, (unidentified), and Henry Nelson. Marcellus Merrill collection.*

*The third floor of the Cushman Building, long thought to be unsafe, made it almost 100 years before succumbing to a heavy spring snow in May 1969. Dan Flenniken photo, courtesy John Tomay Memorial Library*

EPILOGUE

# THE ONCE AND FUTURE SILVER QUEEN

An interstate highway, a burgeoning ski industry, and the uncertainty of Olympic games in Colorado in 1976 merged into a series of concerns for the small town of Georgetown in the 1970s. The 1960 population of 307 included only 3 people more than the official low point of 304 in 1930. While there had been a small "boom" of an additional 100 people in the 1940s, the tiny town slipped close to its all-time low prior to the construction of Interstate 70. The 1970 census showed a strong upswing to 542 residents. Empty houses filled, new homes and businesses began to dot the landscape. Optimism took hold, but so did concerns about the town's ability to balance old and new—how to achieve growth without sacrificing the town, and its wonderful historic properties, to full-scale development.

In December 1968 three relatively new residents of Georgetown sat around the fireplace in the bar of the Alpine Inn drinking cognac: Ronald J. Neely, a part-time bartender and worker at Loveland Ski Area; Cynthia Wadsworth, a schoolteacher, and John Calhoun, a teacher working on his PhD in theater at the University of Colorado. Discussion turned to the need to protect the town from the influx of development that might come from the proposed Olympics and the need to save the architectural integrity of the small town from the more extreme forms of modern architecture starting to show up throughout the state. The three realized that they were too new, too young, and not likely to hold much sway with the local decision-makers. The three toasted "the once and future Silver Queen of the Rockies" and decided to see if there would be interest among some of the better-known locals. A meeting was scheduled with two well-respected businessmen: Bob Bolander, owner of the Shoppe Internationale, and Robert Gibbs, chairman of the town planning commission and owner of the Alpine Inn Restaurant.

After a meeting at the Bolander house, two goals were defined: passage of an ordinance to protect the town's historic properties and formation of a nonprofit organization. The informal group started to gather sample ordinances from different cities around the country. No municipality in Colorado had passed a comprehensive design review ordinance, though several communities did have landmark ordinances to protect specific historic properties or districts within their communities. Discussion in Georgetown centered upon the need to protect the entire community. Attorney Robert Keating wrote the ordinance, and a series of contentious public hearings were held.

During one of the public hearings, local resident Charles Christiansen brought his own soap box, climbed up, and referred to those who supported the ordinance as "Transient empire builders barren of leaves and fruit."[1] With a split vote, the town's police judge, Jim Abbott, cast the deciding vote, "aye," and the town's first Design Review Ordinance took effect in March 1971.

About the same time, Robert Bolander had presided over the incorporation of the Georgetown Society, Inc., on May 1, 1970. The objectives of the organization included four rather lofty goals:

1. To preserve, restore and enhance the historic character, charm, and beauty of Georgetown, Colorado, and its environs.
2. To revitalize, enrich, and extend the cultural environment of Georgetown and the nation through the encouragement, practice, and study of the arts and humanities.
3. To engage in any commercial, theatrical, or industrial enterprise calculated or designed to further the cultural, educational, and historical purposes of

this corporation and to use the proceeds thereof to support or pay for any of its objects and purposes, which may not be materially profitable.

4. To rent, lease, purchase or otherwise acquire, hold, possess, own, mortgage, sell, convey, option, franchise, subdivide, or otherwise dispose of any real or personal property of any kind and description, or any estate therein in any state, territory or district of the United States, or any and all foreign countries.

John Calhoun later remembered the start of the organization, "We talked mostly about keeping traditions, I don't remember us being particularly concerned about Victorian architecture; that was simply a fact. Only later did the Society turn its sights to real estate. I remember sitting at a big town meeting, and someone said 'Sometimes the only way to preserve something is to buy it,' and I guess that has proved true."[2]

About the same time, Ron Neely and local resident and author Polly Chandler traveled to Denver to meet with former mayor James Grafton Rogers. The meeting took place at the University Club over lunch. Neely remembered Rogers as a pleasant, elderly gentleman with a sharp brain. Certainly his background and accomplishments left the young bartender in awe. After about an hour of small talk on a variety of subjects, Neely began to wonder if Roger's brain was beginning to slip; however, "about an hour into the meeting he took all of those unrelated points and tied them up in a little box with a little bow on it and said, 'Here now, what are you going to do with it?' Really what he was talking about was the heart and soul of what this historic district is all about. It really got down to not relying on what he or I thought other people had done, but what were you going to do." Through the years Neely would refer back to Rogers and his meeting in Denver as the inspiration for thinking of projects on a grander and more comprehensive scale—never limiting his ideas to the more traditional role of many historical societies.

Over the next forty years, the Georgetown Society (later changing its name to Historic Georgetown, Inc.) worked to accomplish many of the early ambitions: acquiring five residential buildings to provide visitors with knowledge of the interests and lifestyles of rich and poor alike; building a visitor center at the entrance to town that would serve close to 400,000 people each year, all of whom would leave the building with a better understanding of the

Georgetown–Silver Plume National Historic Landmark District; completing inventories of historic structures; sponsoring publications and lectures on history and architecture; and many other projects within the broad goals outlined in 1970.

Georgetown's treasure trove of nineteenth- and early twentieth-century buildings that had been spared from destruction by the lack of success and "progress" during the first half of the twentieth century became its new mother lode. Carloads of visitors from the Denver metro area, or cross-country travelers buzzing along the interstate highway, soon began to pull off the road to enjoy the town's resources. The fact that the town retained a great deal of its previous atmosphere and charm would become its greatest asset.

Perhaps those visitors stopped for a moment to reflect upon what had transpired in this beautiful setting and to think about the people who had come there before them. Yesterday was not so long ago.

In 1859, as the first prospectors toiled their way up Clear Creek Canyon searching for their golden fortunes, Charles Dickens, in far off England, sat at his desk writing *A Tale of Two Cities*. He opened his story with that now famous sentence, "It was the best of times, it was the worst of times." He could not have summarized better the saga of the community soon to appear where the prospectors searched, Georgetown, Colorado. Today Georgetown retains a great deal of its historic atmosphere, style, and charm, like an era frozen in time. These great assets beckon you, the reader, to come.

# NOTES

## CHAPTER 1: DAYS OF HOPE, YEARS OF DESPAIR

1. Bureau of Labor Statistics, *Effects of Demonetization of Silver on the Industries of Colorado, July 1 to August 31, 1893* (Denver: Smith-Brooks Printing Co., 1893).

2. *Georgetown Courier*, November 7, 1896.

3. *Engineering and Mining Journal*, September 26, 1896.

4. *Engineering and Mining Journal*, December 26, 1896.

5. *Engineering and Mining Journal*, January 23, 1897.

6. *Georgetown Courier*, January 10, 1898.

7. *Engineering and Mining Journal*, January 14, 1899.

8. *Engineering and Mining Journal*, December 25, 1897.

9. Robert G. Pruitt Jr., *Digest of Mining Claim Laws* (Denver: Rocky Mountain Mineral Law Foundation, 1990), 10; Bureau of Labor Statistics, *Effects of Demonetization of Silver on the Industries of Colorado, July 1 to August 31, 1893* (Denver: Smith-Brooks Printing Co., 1893), 31.

10. Harry A. Lee, commissioner, *Report of the State Bureau of Mines* (Denver: State Bureau of Mines, 1901–2; http://www2.cde.state.co.us/artemis/nrserials/nr92001internet/nr92001190102internet.pdf, digitized by the Internet Archive in 2013), 36.

11. Robert G. Pruitt Jr., *Digest*, 1–11; Rossiter W. Raymond, *Mines and Mining in the States and Territories West of the Rocky Mountains* (Washington, DC: Government Printing Office, 1872), 490–502.

12. *Silver Standard* (Silver Plume), July 25, 1896.

13. *Silver Standard* (Silver Plume), February 10, 1900.

14. *Georgetown Courier*, February 10, 1906.

15. Josiah E. Spurr, George H. Garrey, and Sydney H. Ball, *Economic Geology of the Georgetown Quadrangle (Together with the Empire District), Colorado* (Washington, DC: Government Printing Office, 1908).

16. *Georgetown Courier*, March 3, 1906.

17. *Georgetown Courier*, December 1, 1906.

18. *Georgetown Courier*, October 19, 1907.

19. *Georgetown Courier*, April 9, 1910.

## CHAPTER 2: ALL DRESSED UP WITH NO PLACE TO GO

1. *Silver Standard*, July 1, 1893.

2. *Silver Standard*, July 22, 1893.

3. *Silver Standard*, July 29 1893.

4. *Silver Standard*, March 27, 1897.

5. *Silver Standard*, January 25, 1896.

6. *Georgetown Courier*, June 20, 1891.

7. *Georgetown Courier*, June 20, 1891.

8. *Georgetown Courier*, January 2, 1892.

9. *Georgetown Courier*, July 4, 1891.

10. *Georgetown Courier*, April 14, 1893.

11. *Georgetown Courier*, March 11, 1893.

12. *Colorado Miner* (Georgetown), August 18, 1894; this article indicated the shut-off date was scheduled for September 1, but the town council was still discussing the matter.

13. *Colorado Miner* (Georgetown), August 5, 1893.

14. *Georgetown Courier*, August 19, 1893; August 26, 1893.

15. *Georgetown Courier*, May 4, 1895.

16. *Silver Standard*, August 12, 1899; September 9, 1899.

17. *Silver Standard*, September 9, 1899; by this time the plant was definitely utilizing AC power. Further sources will be needed to verify whether the 1893 plant also had AC capability.

18. *Silver Standard*, January 27, 1900.

19. *Georgetown Courier*, November 18, 1899.

20. *Silver Standard*, November 27, 1897.
21. *Silver Standard*, January 28, 1899.
22. *Georgetown Courier*, May 26, 1900.
23. *Georgetown Courier*, October 20, 1900.
24. *Georgetown Courier*, August 4, 1900.
25. *Georgetown Courier*, March 16, 1901.
26. *Silver Standard*, March 22, 1902.
27. *Georgetown Courier*, April 5, 1902.
28. *Georgetown Courier*, September 6, 1902.
29. *Georgetown Courier*, May 17, 1902.
30. *Silver Standard*, June 14, 1902; *Georgetown Courier*, June 14, 1902.
31. *Silver Standard*, April 9, 1904.
32. *Silver Standard*, April 9, 1904.
33. Articles of Incorporation, the Gray's Peak Railroad Company, filed at the office of the Clear Creek County Clerk, May 13, 1904, now on file in the County Archives, Georgetown, Colorado.
34. *Silver Standard*, August 13, 1904.
35. Articles of Incorporation, the Argentine Central Railway Company, filed at the office of the Clear Creek County Clerk, August 11, 1905, now on file in the county Archives, Georgetown, Colorado.
36. *Silver Standard*, November 25, 1905.
37. *Georgetown Courier*, February 3, 1906.
38. *Silver Standard*, July 22, 1905.
39. *Silver Standard*, January 4, 1902.
40. *Georgetown Courier*, March 24, 1906.
41. *Silver Standard*, June 4, 1904.

## CHAPTER 3: THE DOWAGER SILVER QUEEN

1. *Georgetown Courier*, March 30, 1918.
2. *Silver Standard*, July 22, 1905.
3. *Silver Standard*, September 1, 1906.
4. *Georgetown Courier*, December 1, 1906.
5. *Silver Standard*, October 19, 1907.
6. *Silver Standard*, December 13, 1902.
7. *Silver Standard*, October 22, 1904.
8. *Georgetown Courier*, May 12, 1906; January 19, 1907.
9. *Silver Standard*, April 14, 1906.
10. *Silver Standard*, February 2, 1907.

11. *Georgetown Courier*, November 2, 1912.
12. *Georgetown Courier*, November 30, 1912.
13. Published in the *Georgetown Courier*, January 4, 1913.
14. *Georgetown Courier*, October 16, 1915.
15. *Georgetown Courier*, September 2, 1916.
16. *Georgetown Courier*, July 10, 1915.
17. *Georgetown Courier*, March 19, 1910.
18. *Georgetown Courier*, July 18, 1914.
19. *Georgetown Courier*, May 1, 1890.
20. *Georgetown Courier*, April 11, 1891.
21. *Silver Standard*, September 9, 1893.
22. *Silver Standard*, November 11, 1899.
23. *Georgetown Courier*, June 26, 1915.
24. *Georgetown Courier*, April 24, 1915.
25. *Georgetown Courier*, May 22, 1915; May 29, 1915.
26. *Georgetown Courier*, May 22, 1915.
27. *Georgetown Courier*, September 30, 1916.
28. *Georgetown Courier*, April 1, 1916.
29. *Georgetown Courier*, January 27, 1912.
30. *Georgetown Courier*, December 23, 1916.
31. *Georgetown Courier*, June 17, 1911.
32. *Georgetown Courier*, July 3, 1915.
33. *Georgetown Courier*, April 29, 1916.
34. *Georgetown Courier*, April 27, 1907.
35. *Georgetown Courier*, September 14, 1912.

## CHAPTER 4. MINING WHAT THE MOUNTAIN HAS TO OFFER

1. "Industrial Literature and Trade Statistics," *Mining Science* 66, no. 1693 (August 15, 1912): 107.

2. Josiah E. Spurr, George H. Garrey, and Sydney H. Ball, *Economic Geology of the Georgetown Quadrangle (Together with the Empire District), Colorado* (Washington, DC: Government Printing Office, 1908), 214–18.

3. Spurr et al., *Economic Geology*, 23.

4. *Georgetown Courier*, December 13, 1890.

5. *A Brief Treatise on THE INDEPENDENT MINING COMPANY: Its Properties and the Physical and Industrial Conditions to Which They Are Subject* (Georgetown, CO: 1905), 4, 6, 8.

6. Charles Henderson, *Mining in Colorado* (Washington, DC: Government Printing Office, 1926), 32 and 109; Josiah E. Spurr, "Outline of Results," *Economic Geology*, 23.

7. *Silver Standard*, April 27, 1907.

8. *Georgetown Courier*, March 4, 1916.

9. *Georgetown Courier*, December 2, 1916.

10. *Georgetown Courier*, December 23, 1916.

11. *Georgetown Courier*, December 2, 1916.

12. *Silver Standard*, April 27, 1907. *Georgetown Courier*, March 4, October 14 and 28; December 2 and 23, 1916.

13. *Silver Standard*, December 5, 23, 1903; September 24, November 26, 1904; April 29, 1905; July 14, 1906. *Georgetown Courier*, April 4, 1903; June 9, August 25, 1906.

14. *Silver Standard*, October 24, 1903.

15. *Georgetown Courier*, March 25, 1916.

16. *Silver Standard*, July 22, 1905.

## CHAPTER 5. LONG EVENING INTO NIGHT

1. Woodrow Wilson, "Declaration of Neutrality," *Message to Congress*, 63rd Cong., 2d Sess., Senate Doc. No. 566 (Washington, DC: 1914), 3–4.

2. *Georgetown Courier*, September 29, 1917.

3. *Georgetown Courier*, September 29, 1917.

4. *Georgetown Courier*, December 15, 1917.

5. *Georgetown Courier*, March 2, 9, 1917.

6. *Georgetown Courier*, July 26, 1919.

7. *Georgetown Courier*, July 26, 1919; January 24, 31, 1920.

8. Russell Elliott, "Key Pittman," in *The New Encyclopedia of the American West* (New Haven, CT: Yale University Press, 1998), 889–90.

9. *Georgetown Courier*, September 22, 29, November 10, 1917, February 9, 23, March 23, July 13, August 3, 1918. For metal prices, see Charles Henderson, *Mining in Colorado* (Washington, DC: Government Printing Office, 1926).

10. *Engineering and Mining Journal*, March 1, 1919, 397–98.

11. *Georgetown Courier*, May 29, June 5, 19, September 4, November 20, 1920; Henderson, *Mining*, 109.

12. *Georgetown Courier*, August 11, 1917.

13. *Georgetown Courier*, January 5, 1918.

14. *Georgetown Courier*, May 31, 1919.

15. *Georgetown Courier*, August 23, 1919.

16. *Georgetown Courier*, September 11, 1920.
17. *Georgetown Courier*, December 6, 1919.
18. *Georgetown Courier*, September 22, 1917.

## CHAPTER 6. 1917

Epigraph: Veronica M. Elliott, "My Childhood in the 'House on the Hill' in Georgetown, Colorado," *Georgetown Magazine* 1, no. 2 (Fall 1979): 12.

1. *Georgetown Courier*, January 6, 1917.
2. *Georgetown Courier*, August 4, 1917.
3. *Georgetown Courier*, September 8, 1917.
4. *Georgetown Courier*, January 13, 1917.
5. *Georgetown Courier*, March 31, 1917.
6. *Georgetown Courier*, March 30, 1918.
7. *Georgetown Courier*, April 21, 1917; May 19, 1917.
8. *Georgetown Courier*, September 22, 1917; July 13, 1918.
9. *Georgetown Courier*, April 21, 1917.
10. *Georgetown Courier*, March 17, 1917.
11. *Georgetown Courier*, March 31, 1917.
12. *Georgetown Courier*, March 3, 1917.
13. *Georgetown Courier*, February 24, 1917.
14. *Georgetown Courier*, April 14, 1917.
15. "Census List No. 4, 1917, District No. 3, Clear Creek County," and "Annual Report of School District No. Three for the year ending June 30, 1918," both available at Colorado State Archives, Denver, Colorado; scanned copies available at Clear Creek County Archives, Georgetown, Colorado. The 1918 report provides additional information in a format not available for 1917.
16. *Georgetown Courier*, May 12 1917.
17. *Georgetown Courier*, May 19, 1917.
18. *Georgetown Courier*, April 14, 1917.
19. *Georgetown Courier*, May 19, 1917.
20. *Georgetown Courier*, April 21, 1917.
21. *Georgetown Courier*, May 5, 1917.
22. *Georgetown Courier*, June 30, 1917.
23. *Georgetown Courier*, May 26, 1917.
24. *Georgetown Courier*, October 6, 1917.
25. *Georgetown Courier*, October 6, 1917.
26. *Georgetown Courier*, May 19, 1917.

## CHAPTER 7. THEY ROARED AND THEY SPUTTERED

1. *Georgetown Courier*, October 12–December 28, 1918.

2. Gail Collins, *Women in America: 400 Years of Dolls, Drudges, Helpmates, and Heroines* (New York: Harper Collins, 2003), 114.

3. *Mountain States Mineral Age*, August 1922, 7, 11, 16, and 20.

4. *Georgetown Courier*, April 16, 1927.

5. *Georgetown Courier*, March 5, 12, April 16, 1927.

6. *Georgetown Courier*, March 29, 1929.

7. *Georgetown Courier*, October 15, 1921; January 23, 1926; November 5, 1927; December 22, 1928; October 19, 1929; and January 4, 1930.

8. *Georgetown Courier*, March 30, 1929.

9. *Georgetown Courier*, March 30, 1929.

10. *Georgetown Courier*, October 23, 1926.

11. *Georgetown Courier*, April 10, 1926.

12. *Georgetown Courier*, August 23, 1927.

13. *Georgetown Courier*, June 29, 1929.

14. *Georgetown Courier*, July 7, 1928, 1922.

15. *Georgetown Courier*, April 19, 1924.

16. *Georgetown Courier*, November 3, 1928; March 29, 1929.

17. *Georgetown Courier*, March 27, 1926.

18. *Georgetown Courier*, January 3, 1925.

19. *Georgetown Courier*, April 15, 1922; May 10, October 25, November 1, 8, 15, 22, 1924; and January 3, 1925.

20. *Georgetown Courier*, 1922.

21. See, for example, *Georgetown Courier*, September 24, October 1, and November 26, 1921; December 30, 1922; January 1, 1923; April 11, 1925; March 27 and November 13, 1926; April 16, 1927; and December 28, 1929.

22. *Georgetown Courier*, October 1927.

## CHAPTER 8. TIMES ARE GETTIN' HARD BOYS, MONEY'S GETTIN' SCARCE

*Chapter title*: Song lyrics by Lee Hays.

*Epigraph*: Jean Rutherford Duaine, *The Colorado Whoopenhollars: Living a Good Life despite the Great Depression* (Bloomington, IN: Xlibris Corporation, 2009), 19. Jean's recollections of life in Georgetown during the Depression include a treasure trove of stories.

1. Personal recollection of Verona Chappel Zimmerman of Silver Plume, who remembered one of her brothers and George Rowe running the Centennial Mill to produce chicken scratch.

2. *Georgetown Courier*, March 18, 1933.

3. *Georgetown Courier*, November 12, 1932.

4. *Georgetown Courier*, March 12, 1932.

5. *Georgetown Courier*, December 23, 1933.

6. *Georgetown Courier*, November 12, 1932.

7. *Georgetown Courier*, November 26, 1932.

8. *Georgetown Courier*, November 26, 1932.

9. *Georgetown Courier*, December 31, 1932.

10. *Georgetown Courier*, August 19, 1933.

11. *Georgetown Courier*, September 2, 1933.

12. *Georgetown Courier*, November 25, 1933.

13. Clear Creek County Board of County Commissioners Minutes, Book 8, page 53, August 21, 1926, copy available at County Archives, Georgetown, Colorado.

14. *Georgetown Courier*, May 19, 1928.

15. *Georgetown Courier*, April 29, 1933.

16. *Georgetown Courier*, February 10, 1934.

17. *Georgetown Courier*, August 5, 1933.

18. *Georgetown Courier*, February 17, 1939.

19. *Georgetown Courier*, May 19, 1917.

20. *Georgetown Courier*, February 4, 1928.

21. *Georgetown Courier*, August 27, 1932.

22. *Georgetown Courier*, August 27, 1932.

23. *Georgetown Courier*, February 3, 1934.

24. *Georgetown Courier*, March 30, 1929.

25. *Georgetown Courier*, October 31, 1936.

26. *Georgetown Courier*, January 30, 1937.

27. *Georgetown Courier*, September 11, 1937.

28. *Georgetown Courier*, July 22, 1938.

29. *Georgetown Courier*, July 29, 1938.

30. *Georgetown Courier*, July 29, 1938.

31. *Georgetown Courier*, September 2, 1938.

32. *Georgetown Courier*, October 21, 1938. Article includes sketch of the building.

33. *Georgetown Courier*, December 23, 1938.

34. *Georgetown Courier*, December 23, 1938.

35. *Georgetown Courier*, August 4, 1939.

36. *Georgetown Courier*, November 3, 1939.

37. *Georgetown Courier*, January 16, 1937.

38. *Georgetown Courier*, January 23, 1937.

39. Conversations between Christine Bradley and Jim Bennett (Ward, Colorado) grandson of Jonas Bennett, Fall 2009.

40. *Georgetown Courier*, April 28, 1939.

## CHAPTER 9. GEORGETOWN GOES TO WAR

1. *Georgetown Courier*, December 5, 1941.

2. *Georgetown Courier*, December 16, 1941.

3. *Georgetown Courier*, December 26, 1941.

4. *Georgetown Courier*, January 2, 1942.

5. *Georgetown Courier*, February 20, 1942.

6. *Georgetown Courier*, September 10, 1943.

7. *Thomas Hornsby Ferril and the American West* (Golden, CO: Fulcrum Publishing, 1996), 2.

## CHAPTER 10. INTO A NEW ERA

1. The following dates (1952–54) were randomly selected from the last years of the *Georgetown Courier*.

2. Stimson's diary recounts trip west with General Marshall, which included a meeting with Rogers and family in Denver, followed by a trip to Estes Park the next day. While Georgetown is not mentioned in the diary, Rogers referenced the meeting in Georgetown on several occasions. *Denver Post*, February 20, 1955.

3. "Office of Strategic Services: Notes and Comments," by James Grafton Rogers, Box 8, Office of Strategic Services folder, Steven Hart Library, Colorado Historical Society.

4. "In Memoriam," *American Alpine Journal* 18, no. 1 (1972), www.publications.americanalpineclub.org/articles/12197223500/James-Grafton-Rogers-18831971.

5. Thomas F. Troy, *Wartime Washington: The Secret OSS Journal of James Grafton Rogers 1942–1943* (Frederick, MD: University Publications of America, 1987), xxi.

6. Letter from Hugh Wilson, Office of Strategic Services, Washington, DC, January 10, 1944, to Dr. James G. Rogers, Georgetown, Colorado, Box 12, Hugh Wilson folder, James Grafton Rogers Collection, Steven Hart Library, Colorado Historical Society.

7. Stephen H. Hart, "James Grafton Rogers, 1883–1971," memorial written for the Association of the Bar of the City of New York, c. 1971.

8. Minutes, Board of Selectmen, Town of Georgetown, December 4, 1953.

9. Letter from James Grafton Rogers to Hon. Ralph M. Stuck, January 22, 1954, Box 5, Rogers Collection, Steven Hart Library, Colorado Historical Society.

10. *Denver Post*, June 10, 1952.

11. Minutes, Board of Selectmen, Town of Georgetown, February 15, 1954.

12. Letter from James Grafton Rogers, Ex officio Mayor, to Bernard L. Buckley, September 30, 1953, Box 5 Rogers Collection, Steven Hart Library, Colorado Historical Society.

13. Correspondence in Boxes 5 and 7 of the James Grafton Rogers collection, Stephen Hart Library, Colorado Historical Society.

14. Correspondence in Boxes 5 and 7 of the James Grafton Rogers collection, Stephen Hart Library, Colorado Historical Society.

15. Phone conversation between Rosemary Reichwein and Christine Bradley, February 14, 2014.

16. Correspondence with Colorado State Planning Commission, mailed August 10, 1954, Box 5, Rogers Collection, Steven Hart Library, Colorado Historical Society.

17. Miriam W. Adams, "Two Historic Houses Now Preserved as Museums," *Colorado Magazine* 39, no. 2 (April 1962): 107–15.

18. *Georgetown Courier*, June 7, 1956.

19. *Rocky Mountain News*, June 13, 1956.

20. *Rocky Mountain News*, June 6, 1956.

21. *Rocky Mountain News*, June 6, 1956.

22. *Rocky Mountain News*, June 13, 1956.

23. Box 7, James Grafton Rogers collection, Stephen Hart Library, Colorado State Historical Society.

## CHAPTER 11. A TURN TOWARD GROWTH

1. Information from biography of Marjorie Jump on the Colorado Ski Hall of Fame website: www.skimuseum.net (look under "Hall of Fame").

2. Deed found in book 253, page 472 in County Archives, Georgetown. The deed between Mr. Pulliam and the Schussbaumers was dated October 26, 1952, but was not recorded until October 11, 1956, probably after the club completed payment on the property.

3. Discussion with Robert Gibbs, later owner of the Alpine Inn Restaurant and resident of town at the time.

4. Quote handwritten by Mr. Draper in one of his scrapbooks now in the possession of Coralue Anderson of Georgetown, Colorado.

5. *San Francisco Examiner*, June 21, 1946, clipped and pasted in Draper scrapbooks mentioned above.

6. *Rocky Mountain News*, April 7, 1946, in Draper scrapbooks.

7. *Denver Post*, April 30, 1946, clipping from Draper scrapbooks.

8. Photo labeled "Squeak Morris, Craig Abrahamson, and Ben," in Draper scrapbooks.

9. Program entitled "1946, Sixth Annual Western Folklore Conference, July 11–13, University of Denver, Denver, Colorado," in Draper scrapbooks.

10. *Denver Post*, July 2, 1946, clipping from Draper scrapbooks.

11. *Rocky Mountain News*, August 24, 1946.

12. Charles B. Hosmer Jr., *Preservation Comes of Age: From Williamsburg to the National Trust, 1926–1949* (Charlottesville: University Press of Virginia, 1981), vol. 1: 373–75.

13. Hosmer, *Preservation Comes of Age*, 375.

14. Letter from James Grafton Rogers to Henry K. Anderson, November 22, 1954, Box 5, folder 5, James Grafton Rogers Collection, History Colorado.

15. "The National Survey of Historic Sites and Buildings, Theme XV, Westward Expansion and the Extension of the National Boundaries to the Pacific, 1830–1898; Overland Migrations West of the Mississippi" (Washington, DC: United States Department of the Interior, National Park Service, 1959).

16. James Grafton Rogers, "Report on the Sites and Structures of Historical Interest in the State of Colorado," 1954, Box 10, James Grafton Rogers Collection, History Colorado.

17. Rogers, "Report on the Sites and Structures," 6.

18. Letter from Jared B. Morse, architect, to James Grafton Rogers November 11, 1958, Box 5, James Grafton Rogers Collection, History Colorado.

19. "How the Railroad Came Back to Silver Plume," *Clear Creek Courant*, October 12, 1973.

20. Letter to the editor, Marcellus Merrill, *Clear Creek Courant*, July 14, 1982.

21. *Clear Creek Courant*, October 12, 1973.

## EPILOGUE: THE ONCE AND FUTURE SILVER QUEEN

1. Conversation between Christine Bradley and Ronald J. Neely, December 2001.

2. Aysha Griffin, "Historic Georgetown, Inc., 1970–1995," unpublished manuscript.

# INDEX

*Page numbers in italics indicate illustrations.*